POLYMER SCIENCE and ENGINEERING

The Shifting Research Frontiers

Committee on Polymer Science and Engineering

Board on Chemical Sciences and Technology

Commission on Physical Sciences, Mathematics, and Applications

National Research Council

National Academy Press
Washington, D.C. 1994

National Academy Press • 2101 Constitution Avenue, N.W. • Washington, D.C. 20418

NOTICE: The project that is the subject of this report was approved by the Governing Board of the National Research Council, whose members are drawn from the councils of the National Academy of Sciences, the National Academy of Engineering, and the Institute of Medicine. The members of the committee responsible for the report were chosen for their special competences and with regard for appropriate balance.

This report has been reviewed by a group other than the authors according to procedures approved by a Report Review Committee consisting of members of the National Academy of Sciences, the National Academy of Engineering, and the Institute of Medicine.

Support for this project was provided by the National Science Foundation, Defense Advanced Research Projects Agency, Department of Army, Department of Energy (Grant No. DE-FG05-92ER-45478), E.I. du Pont de Nemours & Co., Air Force Office of Scientific Research, Department of the Navy, and the Basic Science Fund of the National Academy of Sciences, whose contributors include the AT&T Foundation, Atlantic Richfield Foundation, BP America, Dow Chemical Company, E.I. du Pont de Nemours and Company, IBM Corporation, Merck and Company, Inc., Monsanto Company, and Shell Oil Companies Foundation.

Library of Congress Cataloging-in-Publication Data
Polymer science and engineering : the shifting research frontiers /
Committee on Polymer Science and Engineering, Board on Chemical Sciences and Technology, Commission on Physical Sciences, Mathematics, and Applications, National Research Council.
p. cm.
Includes bibliographical references and index.
ISBN 0-309-04998-9
1. Polymers—Research. I. National Research Council (U.S.). Committee on Polymer Science and Engineering.
QD281.P6P635 1994
668.9—dc20 94-21613
CIP

Printed in the United States of America

COMMITTEE ON POLYMER SCIENCE AND ENGINEERING

RICHARD S. STEIN, University of Massachusetts at Amherst, *Chair*
LESTER C. KROGH, 3M (retired), *Vice Chair*
DOTSEVI Y. SOGAH, Cornell University, *Vice Chair*
C. JEFFREY BRINKER, Sandia National Laboratory
KENNETH A. DILL, University of California at San Francisco
ROBERT H. GRUBBS, California Institute of Technology
EDWARD J. KRAMER, Cornell University
SONJA KRAUSE, Rensselaer Polytechnic Institute
JAMES E. MARK, University of Cincinnati
DAVID W. McCALL, AT&T Bell Laboratories (retired)
JAMES E. McGRATH, Virginia Polytechnic Institute and State University
JAMES E. NOTTKE, E.I. du Pont de Nemours & Co.
DONALD R. PAUL, University of Texas at Austin
S. ELAINE PETRIE, Eastman Kodak Company (retired)
DAVID A. TIRRELL, University of Massachusetts at Amherst
C. GRANT WILLSON, University of Texas at Austin

Staff

TAMAE MAEDA WONG, Study Director
MARIA P. JONES, Project Assistant

COMMISSION ON PHYSICAL SCIENCES, MATHEMATICS, AND APPLICATIONS

The National Academy of Sciences is a private, nonprofit, self-perpetuating society of distinguished scholars engaged in scientific and engineering research, dedicated to the furtherance of science and technology and to their use for the general welfare. Upon the authority of the charter granted to it by the Congress in 1863, the Academy has a mandate that requires it to advise the federal government on scientific and technical matters. Dr. Bruce Alberts is president of the National Academy of Sciences.

The National Academy of Engineering was established in 1964, under the charter of the National Academy of Sciences, as a parallel organization of outstanding engineers. It is autonomous in its administration and in the selection of its members, sharing with the National Academy of Sciences the responsibility for advising the federal government. The National Academy of Engineering also sponsors engineering programs aimed at meeting national needs, encourages education and research, and recognizes the superior achievements of engineers. Dr. Robert M. White is president of the National Academy of Engineering.

The Institute of Medicine was established in 1970 by the National Academy of Sciences to secure the services of eminent members of appropriate professions in the examination of policy matters pertaining to the health of the public. The Institute acts under the responsibility given to the National Academy of Sciences by its congressional charter to be an adviser to the federal government and, upon its own initiative, to identify issues of medical care, research, and education. Dr. Kenneth I. Shine is president of the Institute of Medicine.

The National Research Council was organized by the National Academy of Sciences in 1916 to associate the broad community of science and technology with the Academy's purposes of furthering knowledge and advising the federal government. Functioning in accordance with general policies determined by the Academy, the Council has become the principal operating agency of both the National Academy of Sciences and the National Academy of Engineering in providing services to the government, the public, and the scientific and engineering communities. The Council is administered jointly by both Academies and the Institute of Medicine. Dr. Bruce Alberts and Dr. Robert M. White are chairman and vice chairman, respectively, of the National Research Council.

Preface

The last decade has produced dramatic changes in national scientific and economic issues. Environmental goals have led to new standards, and the end of the Cold War has shifted national priorities from military to economic security. These changes have had direct effects on priorities for research and development in both the public and the private sector. The growing economic and industrial sophistication of other countries also presents new challenges for our economic stability. If the United States is to maintain its leadership role and ability to compete in the global market, we must clearly understand the frontiers of research in order to plan for the future.

In 1992, a committee was established by the Board on Chemical Sciences and Technology (BCST) of the National Research Council (NRC) to assess the research frontiers in polymer science and engineering. Given the scientific advancements in the field since the publication of the 1981 NRC report, *Polymer Science and Engineering: Challenges, Needs, and Opportunities* (National Academy Press, Washington, D.C.), it was clear that another look at polymer science and engineering was in order. The goals were to examine the recent advances in research and to identify new thrusts in the context of current and long-term national needs and concerns. The committee was charged to

- Identify ways that polymer research contributes to the solution of important national issues;
- Encourage the scientific and technological community to give increased attention to advancing the frontiers of research and education; and

• Recommend priorities to enable administrators, policymakers, and funding agencies to optimize the use of limited resources.

The study began with a workshop held in Washington, D.C., on March 26 and 27, 1992, at which invited specialists presented their views of their fields, outstanding scientific challenges, and areas that they thought should be emphasized. In addition, the committee surveyed other scientists, engineers, and industrialists by mail to obtain their views. Discussions among committee members were conducted over four meetings during 1992 and 1993, and consensus on the research priorities as identified in the recommendations was reached by the committee members. The broad state-of-the-art report that resulted is aimed at a diverse audience. For polymer research specialists, it offers a summary of current activities in research and commercial technologies. For investigators in other branches of materials science and fields applying polymers such as biomedicine and electronics, it reviews new directions in polymer research, emphasizing important interdisciplinary opportunites. Finally, for leaders in science policy and funding, this report presents topics chosen for their importance to society and delineates priority areas in polymer science and engineering.

The committee's chair is indebted to its members for the many hours that this able and conscientious group devoted to this effort. In particular, David W. McCall is recognized for his dedication and insight in editing the report into its final form. The efforts of BCST staff members Tamae Maeda Wong, Douglas J. Raber, and Maria P. Jones were also of major importance. The contributions of Douglas L. Smith of the California Institute of Technology, who edited the material for the vignettes, are very much appreciated. The appendix lists additional participants and writing contributors. All of us understood the importance of the task, whose results we hope will benefit the polymer community and the nation.

Richard S. Stein, *Chair*
Committee on Polymer Science and Engineering

Contents

POLYMER SCIENCE and ENGINEERING

The Shifting Research Frontiers

Summary and Recommendations

A revolution has taken place over the last 50 years in the field of synthetic polymers, whose applications have rapidly permeated most aspects of our daily lives. The scientific and technical advances, and the consequent business successes, have been steady and abundant. However, in this post-Cold War era of growing international economic competitiveness, it is clear that the United States must plan wisely if it is to participate fully in the future of this dynamic field. The investment will allow the United States to continue to enjoy the many societal benefits that will flow from research and development in polymer science and engineering.

To help provide a basis for such planning, the National Research Council's (NRC) Board on Chemical Sciences and Technology asked the committee responsible for this report to conduct a comprehensive assessment of research opportunities in polymer science and engineering and to point out the field's contributions to national issues. The assignment included definition of research frontiers, identification of national research and educational challenges, and delineation of the corresponding funding priorities needed for planning by funding agencies. The report builds on two earlier NRC reports, *Polymer Science and Engineering: Challenges, Needs, and Opportunities* (NRC, 1981) and *Materials Science and Engineering for the 1990s* (NRC, 1989).

POLYMER SCIENCE AND ENGINEERING: RELEVANCE AND OPPORTUNITIES

Polymers are molecules that contain many atoms, typically tens of thousands to millions. While many polymers occur naturally as products of biological processes, synthetic polymers are made by chemical processes that combine many small units, called monomers, together in chains, branched chains, or more complicated geometries. Starch, cellulose, proteins, and DNA are examples of natural polymers, while nylon, Teflon®, and polyethylene are examples of the synthetic variety. Both classes possess a number of highly useful properties that are as much a consequence of the large size of these molecules as of their chemical composition. Although most synthetic polymers are organic, that is, they contain carbon as an essential element along their chains, other important polymers, such as silicones, are based on noncarbon elements.

The rapid pace of advances in polymers, with only a few decades separating their first commercial development from their present pervasive use, has been remarkable. Synthetic polymers are so well integrated into the fabric of society that we take little notice of our dependence on them. This is truly the polymer age! Society benefits across the board—in health, medicine, clothing, transportation, housing, defense, energy, electronics, employment, and trade. Without a doubt, synthetic polymers have large impacts on our lives.

Although progress to date in polymer science and engineering can be considered revolutionary, opportunities are abundant for creating new polymeric materials and modifying existing polymers for new applications. Scientific understanding is now replacing empiricism, and polymeric materials can be designed on the molecular scale to meet the ever more demanding needs of advanced technology. The possible control of synthetic processes by biological systems is promising as a means of perfecting structures. New catalysts offer the opportunity to make new materials with useful properties, and the design of new specialty polymers with high-value-added applications is an area of rapidly increasing emphasis. Theory, based in part on the availability of high-speed computing, offers new understanding and aids in the development of improved techniques for preparing polymers as well as predicting their properties. Analytical methods, including an array of new microscopic techniques particularly suited to polymers, have been developed recently and promise to work hand-in-hand with theoretical advances to provide a rational approach to developing new polymers and polymer products. The field of polymer science and engineering therefore shows no sign of diminished vigor, assuring new applications in medicine, biotechnology, electronics, and communications that will multiply the investment in research many times over in the next few decades.

FACTORS AFFECTING U.S. STRENGTH IN POLYMERS

The U.S. Response to Global Competition

Changing world conditions and shifting national priorities have necessitated a reexamination of U.S. research and development activities. What are the potential benefits of polymer science and engineering? What can the United States do to maintain its strength in this promising but internationally competitive area? Can we ensure that our citizens will benefit from developments over the long term as well as in the immediate future? More specific questions that need to be addressed include the following:

- Will government policy encourage a state of health in the polymer industry?
- What level of research and development spending is necessary for the long term to enable effective competition with other nations?
- How can funding for university, industry, and government laboratories facilitate the development of new technologies and products that will benefit society?
- How can state-of-the-art industrial infrastructure be maintained for processing and production equipment?
- How should production, distribution, use, and disposal of polymeric materials be managed to ensure protection of the environment and the health of the public?

Current Conditions and Trends in Research

Federal support for polymer research has always been modest in comparison with all funding for advanced materials and processing. The federal materials research and development budget request for 1993 was $1,821 million (FCCSET, 1992). The polymer science and engineering portion was only $93 million (5%), although some fraction of the budgets for other categories includes support for polymer research. The basic research segment for polymers from the National Science Foundation amounted to only $23 million in 1993.

Industrial support for polymer science and engineering research has been very strong over the decades, but that support appears to be ebbing quickly as corporations retreat from long-range research to research aimed at near-term product introduction or modification. The principal U.S. competitors, Germany and Japan, will contest U.S. leadership in research (NSF, 1992) and thus challenge one of the few remaining areas of U.S. trade that posts a positive balance. Meeting this challenge will depend on maintaining a solid research effort.

Among the worldwide trends in the polymer industry is a shift in R&D focus from commodity plastics, produced in massive quantities, to engineering plastics that have superior properties but are produced in lower volumes. In recent years, the emphasis has been on specialty polymers that are expensive yet have specific properties that confer high value, for example, medical prostheses such as replacement tendons and hip cups, or flexible light-emitting diodes. Although the specialty market is still emerging, it is clear that it will be research intensive and highly competitive. Failure to support this area now could limit U.S. participation in the benefits of applications. Beyond funding, there is a need to nurture relationships between polymer researchers and practitioners of medical, biological, electronic, and other fields of application. These researcher-practitioner relationships are poorly established in the United States at this time, and federal funding policies could enhance such interactions, for example, by encouraging joint grants to foster collaboration.

Another concern arises from the interdisciplinary nature of polymer science and engineering, and the lack of the field's integration into most university curricula. Isolated faculty members specializing in polymer research can be found in many chemisty, chemical engineering, and materials science and engineering departments, yet at many universities the barriers between departments prevent effective interdisciplinary collaboration in polymer research and teaching. Most students in technical programs currently receive little training in polymer science and engineering, in spite of the fact that in many of these fields more than half of the students will eventually pursue careers in which they will be centrally involved with polymers. As achievements in polymer research become more accessible to faculty in traditional fields, polymer science may be expected to become part of the core education in science. In the interim, substantial benefits would result from an emphasis in academic programs on forming teams across groups or disciplines to carry out interdisciplinary work at the frontiers of polymer science and engineering.

The comprehensive understanding of international and national developments and planning for constructive change involving government, industry, and academia, therefore, are central to our nation's full enjoyment of the many benefits of polymer science and engineering research.

LOOKING AHEAD: FINDINGS, CONCLUSIONS, AND RECOMMENDATIONS

Underlying Principles

The committee's conclusions and recommendations are based on the following tenets. First, it is essential that a strong basic research community be sustained. Short-term or product research will not suffice to prepare for future challenges. Second, strong ties between basic research and applications need to be maintained in order to reap maximum commercial benefits. Thus, industry must continue polymer research and maintain communication with academic scientists and engineers. Finally, the committee believes that success in developing the next generation of commercial polymers will depend on strong and extensive collaborative research at the interfaces between polymers and other areas of science and engineering.

Summarized below are the committee's main findings and recommendations resulting from its deliberations. Additional conclusions and recommendations are given in the main text of the report.

Research Balance for Long-term National Well-being

Findings and Conclusions: Polymeric materials production, a large, diverse industry in which the United States has been a leader for several decades, accounted for revenues of over $100B, employment of over 170,000, and a positive trade balance of about $6B in 1992. Polymers have broadly penetrated the materials markets at the commodity, engineering, and high-technology specialty levels. Examples include automobile and airframe components, fibers and fabrics, rubber products of all kinds, and packaging and structural plastics.

Observed trends, however, have raised concerns that the era of leadership and positive contribution to the U.S. economy is in danger of coming to an end. Industrial research funding generally reported indicates a 7 percent increase in 1992 over 1991 (*Business Week*, 1993), but other surveys and economic indicators are in conflict with such an optimistic analysis. Particularly worrisome are recent organizational changes in industry and the shortening of research horizons to focus on improving existing products and on bringing products to market more rapidly at the expense of research directed to achieving basic understanding and breakthroughs in new materials.

Recommendation 1: To ensure a basis for future success in the U.S. polymer industry and concomitant long-term social and economic benefits for the nation, the committee recommends a broad reassessment of the current balance between research and development in polymer science and engineering. Consideration should be given to the following:

- Maintenance of corporate research groups that have a viable nucleus of highly qualified specialists, to enable corporations to take advantage of continuing advances and breakthroughs;
- Development of government policies and legislation that encourage achievement of long-term, rather than just short-term, goals by industry and that stimulate industry and economic well-being;
- Funding by government of programs that encourage industry collaboration with academia and with national laboratories; and
- Increased funding for polymer research that reflects the significance of polymers as a key element in the current materials initiatives.

Increased Interaction Between Polymer Researchers and Practitioners: Need for Interdisciplinary Approaches and Team Efforts

Findings and Conclusions: As the field of polymer science and engineering and the problems it addresses have grown larger and more complex, the need has increased for an integrated approach to achieving improvements in such key areas as manufacturing, transportation, energy, housing, medicine, information and communications, and defense.

Recommendation 2: To expand the technology base and secure for the nation the benefits of research at the frontiers, the committee recommends that researchers and funders give high priority to work that strengthens the interface between polymer science and engineering and the many areas, such as medicine and electronics, in which it is applied.

Attention to Supporting High-Priority Frontier Research Areas

Findings and Conclusions: Scientific and technological progress during the 1980s has created rich opportunities for research in polymer science and engineering. The potential is great for future developments that will strongly contribute to areas of national concern.

Recommendation 3: The committee recommends high priority for support of research and education in the following frontier areas:

- Interdisciplinary investigations of polymer surfaces and interfaces, including studies to increase understanding of chemical reactions that take place at surfaces and research to enable making smaller structures, thin films, and nanophase materials that have the same scale as the morphological features of polymers;
- Synthesis of new polymers and polymeric materials, including methods that precisely control the structure of polymers, biosynthesis, catalysis, and environmentally benign synthesis;
- New methods for processing and manufacturing materials, including computer-assisted design of processing and on-line process control;
- Characterization of polymers and development of new methods (including, new techniques in microscopy and magnetic resonance imaging) to aid visualization and understanding of polymer properties; and
- Theory including modeling, statistical mechanics, and molecular dynamics studies that take advantage of the unprecedented computing power available now and in the foreseeable future. These methods will allow modeling of complex processing operations, prediction of mechanical behavior and failure, calculation of thermodynamic and time-dependent properties, and the design of molecules capable of highly specific molecular recognition.

Further Study of Environmental Issues Related to Materials

Findings and Conclusions: Growing national concern about protecting Earth's resources and the need for sound scientific understanding as a foundation for nonadversarial, constructive environmental legislation present both challenges and opportunities for polymer science and engineering. Increasingly, technical progress in polymer science and engineering is driven by environmental considerations. For example, the environmental impact of using polymeric materials depends in part on how they are disposed of at the end of their useful life. Clearly, no single pathway of disposal is optimal for all materials, and researchers are looking at several alternatives: direct recycling, degradation, and incineration with heat recovery as opposed to landfill disposal. Efforts also are being made to minimize waste by-products during manufacture and to extend the useful life of materials by improving their properties. Emissions reduction has become a major goal of virtually all polymer-producing enterprises. Opportunities exist for polymer scientists and engineers to contribute to the development of more environmentally benign products and processes. However, understanding and dealing effectively with the difficult issues concerning environmental impact will require an integrated approach drawing on the strengths of polymer science and engineering, economics, and policy analysis.

Recommendation 4: The committee recommends that an independent committee at the national level be appointed to accomplish the following:

- Analysis of the environmental issues posed by materials, including polymers; and
- Scientific, engineering, and economic analyses of polymeric materials production, processing, use, recycling, and end-use disposal as a guide to environmental policymaking.

Encouragement of Active Collaboration Across Subfields of Materials Science and Engineering

Findings and Conclusions: Interdisciplinary polymer science and engineering research in nontraditional areas and with nontraditional partners will have maximum impact on developments in science and technology and their contribution to ensuring U.S. economic strength and international competitiveness. Yet the fields concerned with broad classes of materials, such as metals, ceramics, electronic materials, biological materials, and polymers, continue to be quite separate in terms of professional societies, academic disciplines, publication media, and industrial organizations. Moreover, the differing technical languages and cultures of the subfields have complicated efforts to establish interactions and collaborations. Even within the polymer area, which is characterized by breadth and diversity, communication can be limited by the insularity of the subfields.

Currently, fragmentation is particularly evident at the interfaces between polymer science and other technical areas. Closer ties between the polymer research community and, for example, groups studying medical and biological materials or those focusing on structural composites and electronic and optical materials could provide important synergies. To achieve progress, bridges must be built to link diverse disciplines and fields. All materials fields would benefit from closer relationships and better communication.

Recommendation 5: To enhance progress in polymer science and engineering through increased collaboration across the subfields of materials science and engineering, the committee recommends the following:

- Initiation of efforts to better integrate the field of polymer science and engineering both internally among polymer subdisciplines and externally with other materials subfields, including, for example, establishment of interdisciplinary programs by academia, initiation of cooperative programming by professional societies, funding of synergistic cross-boundary projects by government agencies, and structural reorganization by industrial laboratories to maximize cooperation among the various materials subfields;
- Adoption by funding agencies of a broader definition of polymer science and engineering in order to foster advances in interdisciplinary research and education; and
- Efforts to take advantage of research opportunities at the interfaces between polymer science and engineering and other materials areas.

REFERENCES

Business Week. 1993. "R&D Scoreboard: In the Labs, the Fight to Spend Less, Get More." June 28, pp. 102-127.

Federal Coordinating Council for Science, Engineering, and Technology (FCCSET). 1992. *Advanced Materials and Processing: The Fiscal Year 1993 Program*. A report by the Office of Science and Technology Policy, FCCSET Committee on Industry and Technology. (Available from Committee on Industry and Technology/COMAT, c/o National Institute of Standards and Technology, Rm. B309, Materials Building, Gaithersburg, MD 20899.)

National Research Council (NRC). 1981. *Polymer Science and Engineering: Challenges, Needs, and Opportunities*. Washington, D.C.: National Academy Press.

National Research Council (NRC). 1989. *Materials Science and Engineering for the 1990s: Maintaining Competitiveness in the Age of Materials*. Washington, D.C.: National Academy Press.

National Science Foundation (NSF). 1992. *National Patterns of R&D Resources: 1992*. NSF 92-330. Washington, D.C.: U.S. Government Printing Office.

1

National Issues

This chapter discusses some of the direct societal benefits that derive from the field of polymer science and engineering and illustrates how this specialized area can contribute to the solution of some of the pressing problems now facing the United States. In terms of commerce, polymers constitute about one-third of the U.S. chemical industry, and the same fraction describes the contribution of polymers to the chemical industry's favorable balance of payments. Polymers have become an essential and ubiquitous part of our lives: clothing, car parts, boat hulls, tennis racquets, aircraft frames, telephones, computers, human body prostheses, and numerous other material goods that we take for granted are composed largely of polymers. This widespread use is a gratifying confirmation of the success of the polymer science and engineering community, but it carries with it the need to continually focus on maintaining the research and development effort that will keep the United States at the forefront of this dynamic field.

The word "polymer" is derived from the Greek roots "poly" and "mer," which mean "many parts." Polymeric substances are composed of many chemical units called monomers, which are joined together into large molecular chains consisting of thousands of atoms. The monomers can be connected in linear chains, branched chains, or more complicated structures, each variety yielding interesting and useful properties. Most polymers are derived from petroleum and are based on the chemistry of carbon, although some polymers have non-carbon-based compounds (e.g., silicones have backbones composed of alternating silicon and oxygen atoms, with organic—carbon-containing—side groups attached to the silicon atoms).

The large size of polymer molecules is the key factor that distinguishes polymers as a class of materials. Hydrocarbons, ranging from natural gas, to gasoline, to paraffin wax, to polyethylene, illustrate the effect of molecular size. As a consequence of molecular size alone, an enormous change in physical properties is observed—from gases, to fluids, to waxy solids, to tough and durable building materials of many uses.

Cotton, linen, hemp, wool, and natural rubber are examples of polymers that occur in nature, while synthetic (or man-made) polymers include nylon, epoxies, polyethylene, Plexiglas, Styrofoam, Kevlar®, and Teflon®. Unless otherwise indicated, the term "polymer" refers to synthetic polymers throughout this report. The term "plastics" is often used as a synonym for synthetic polymers, although some synthetic polymers are not plastic in the sense of being permanently deformable.

Not surprisingly, the applications of polymers vary widely. Given their role in such vital national concerns as economic competitiveness, transportation and energy, and defense, it is necessary to keep in view the opportunities and needs associated with polymer science and engineering, so that appropriate efforts can be made to sustain U.S. well-being and leadership. Currently, however, there are many critical issues, both financial and political. For example, industrial competition has led many corporations to cut back on long-range research in favor of more immediate bottom-line results. University research budgets are under pressure from many directions. Federal funding, in which military research and development has always been a major factor in the United States (as distinguished from other industrialized nations), is in retreat with the end of the Cold War. The need for continued investment for future benefits has never been so great, nor the prospects for such investment more threatened.

The extent to which contributions from polymer science and engineering can be brought to bear on pressing national problems depends in part on the extent to which the field can maintain its strength. Continuing strength will depend on several factors that need to be addressed:

- Will government policy encourage a state of health in the polymer industry?
- What level of research and development spending is necessary for the long term to enable effective competition with other nations?
- How can funding for university, industry, and government laboratories facilitate the development of new technologies and products that will benefit society?
- How can state-of-the-art industrial infrastructure be maintained for processing and production equipment?
- How should production, distribution, use, and disposal of polymeric materials be managed to ensure protection of the environment and the health of the public?

RESEARCH FUNDING

Research funding for polymer science and engineering is not reported separately in most surveys. In the absence of such figures, a look at total U.S. research and development expenditures is useful. As shown in Figure 1.1, growth in expeditures for research and development from 1975 to 1985 has been fol-

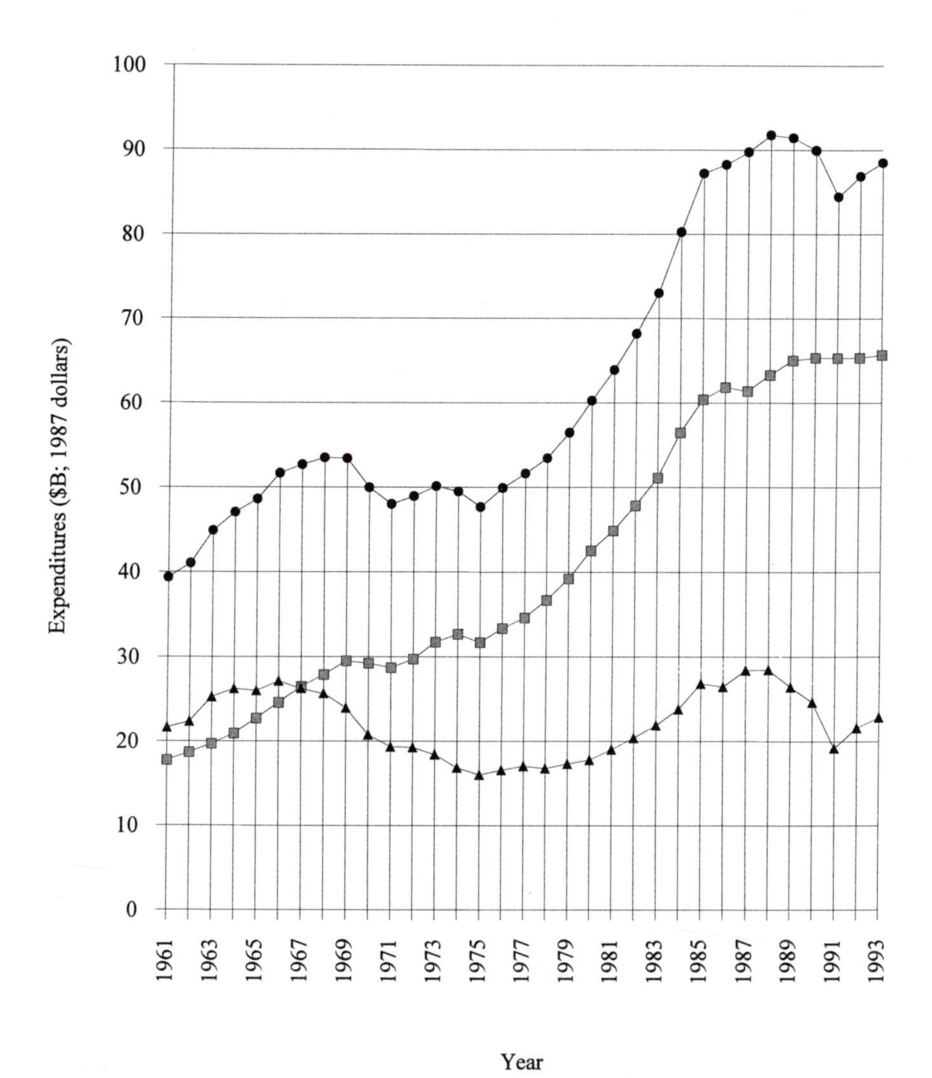

FIGURE 1.1 Expenditures for U.S. industrial R&D, 1961 to 1993. Funded by industry, shaded squares; funded by government, solid triangles; total, solid circles. SOURCE: Data from National Science Board (1993).

lowed by several years of almost level funding. Note that Figure 1.1 is plotted in 1987 dollars; that is, inflation has been factored out. The industry-funded fraction has remained consistently strong, even during the recent recessionary period. It should be emphasized that more than 70 percent of the funded research and development is "development." Applied and basic industrial research expenditures are given in Figure 1.2. The trends are similar to those for overall industrial research and development, and the fraction of the total is 20 to 25 percent for applied research and 4 to 5 percent for basic research. The ratios among the components have remained remarkably constant over two decades. Figure 1.3 exhibits data for industry-funded research and development in universities and nonprofit institutions. University programs are likely to fall into the basic research or applied research categories. The continued growth of total U.S. research and development in difficult times is heartening.

Recent data show that U.S. expenditures in 1992 for research and development were up by 7 percent over 1991 (*Business Week*, 1993a) and that this strength covered a broad range of industries. Only telecommunications, metals, and fuels posted declines. Foreign expenditures for research grew by 8 percent. Details for a selected list of U.S. producers and users of polymers are given in Table 1.1. The budget figures in Table 1.2 show that funding for polymer research and development is modest (FCCSET, 1992).

Although the broad-brush statistics suggest that U.S. research and development funding is strong, closer analysis provides a darker picture. The National Science Board (NSB) has recently published a major study of U.S. industrial science and technology funding, and its conclusions are much less optimistic than those suggested by the foregoing figures (NSB, 1992). The issues are complex, and the reader is referred to the NSB document. The NSB panel concluded that

- U.S. industrial research and development is in trouble,
- Spending is lagging,
- Expenditures are not well allocated, and
- Research and development is not utilized effectively.

The committee also finds it difficult to reconcile the positive statistical evidence for research and development support by industry with committee members' direct knowledge of layoffs and large reductions in the research staffs of industrial laboratories in the early 1990s. Virtually all of the top spenders in the polymer research areas (and this includes most of the major industrial research organizations) have offered programs to reduce staff. These programs have been humanely framed as early retirement opportunities, but the result is fewer researchers. At the other end of the scale, hiring is being curtailed by industry, academia, and government. New Ph.D.s are having a difficult time finding employment. These facts seem inconsistent with reported increases in research and development budgets. Although the results of statistical surveys can lag

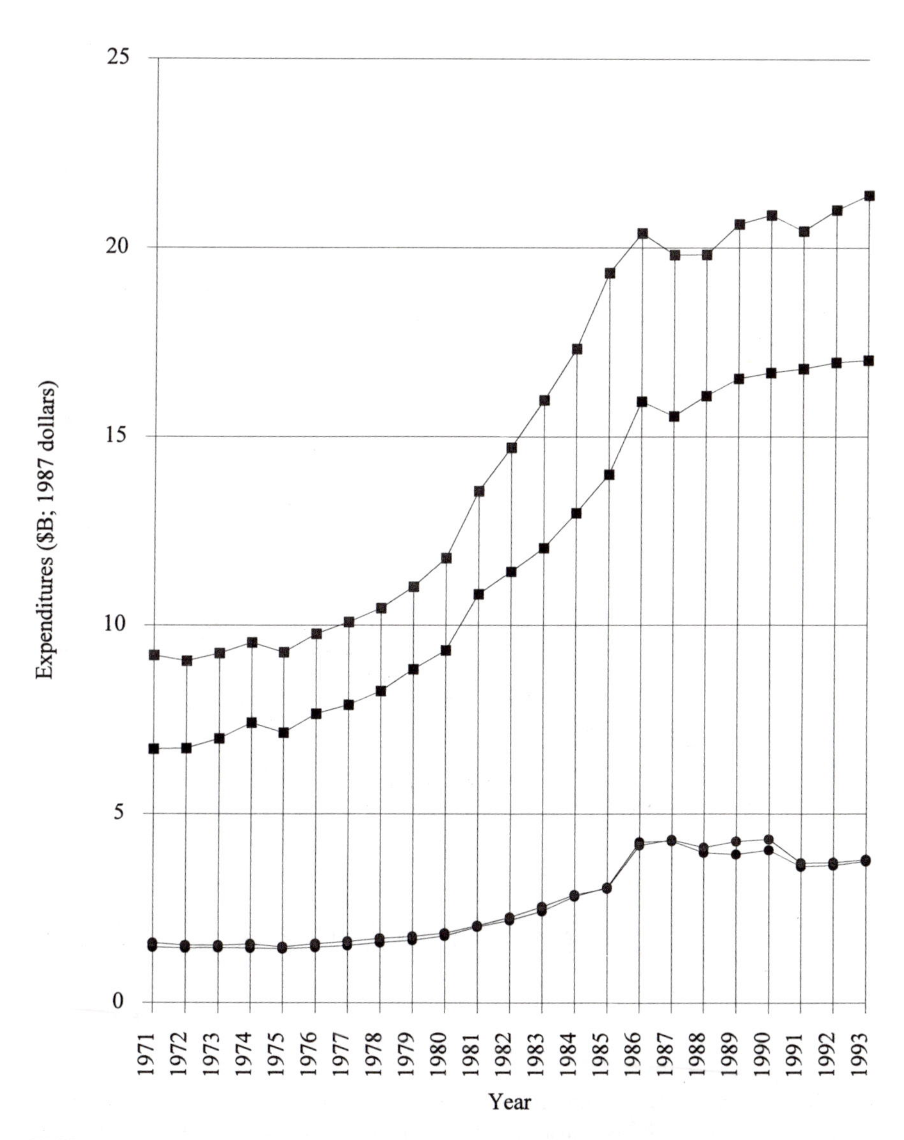

FIGURE 1.2 Applied and basic industrial research expenditures, 1971 to 1993. Applied research—industry funded, solid squares; applied research—industry performed, shaded squares; basic research—industry funded, solid circles; basic research—industry performed, shaded circles. SOURCE: Data from National Science Board (1993).

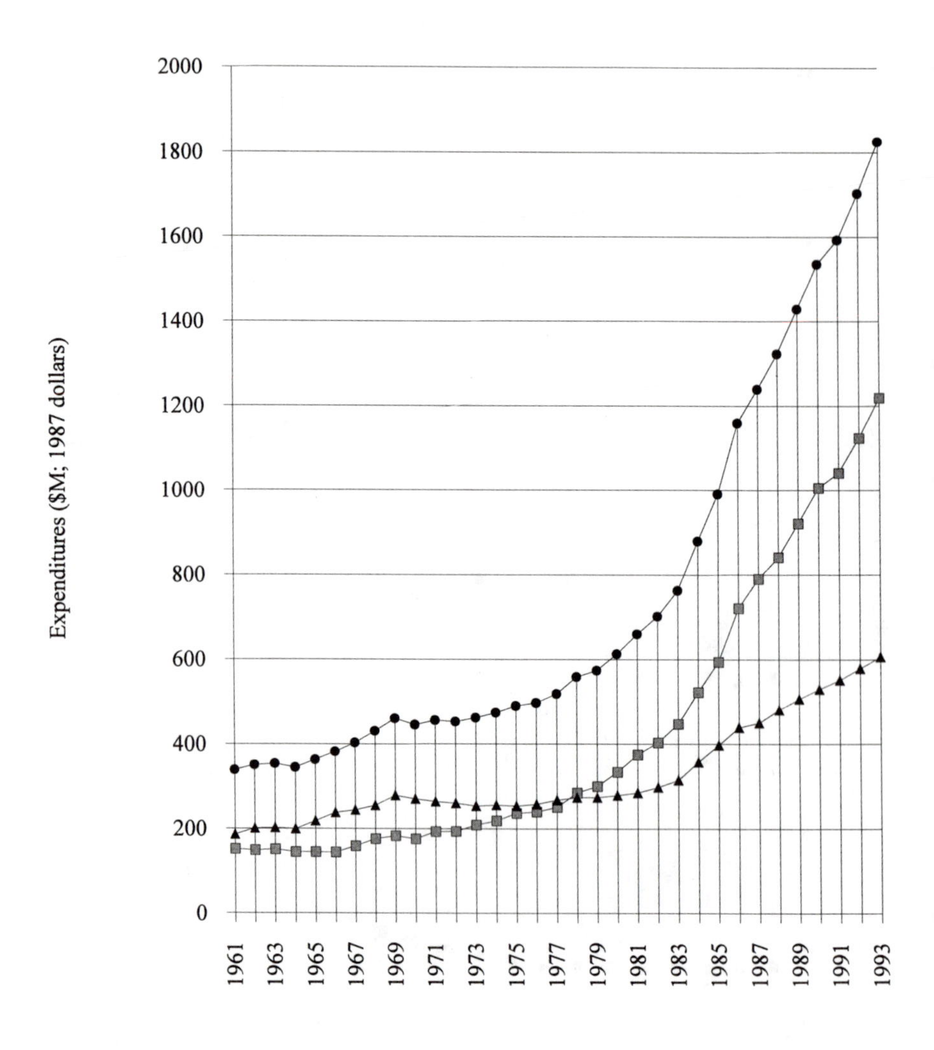

FIGURE 1.3 University and nonprofit R&D funded by industry, 1961 to 1993. University funded by industry, shaded squares; nonprofit funded by industry, solid triangles; total, solid circles. SOURCE: Data from National Science Board (1993).

TABLE 1.1 U.S. Industry Sales and Expenditures for R&D in 1992, Selected Companies

Company	Sales ($B)	Expenditures for R&D ($B)	Percentage Change in Expenditures for R&D from 1991 to 1992
Allied	12	0.3	–16
Dow	19	1.3	+11
du Pont	37	1.3	–2
Eastman Kodak	20	1.6	+6
Exxon	103	0.6	–8
General Electric	56	1.4	+13
Goodyear	12	0.3	–1
Grace	5.5	0.15	+1
3M	14	1.0	+10
Monsanto	7.8	0.65	+7
Phillips	12	0.1	–19
Rohm & Haas	3.1	0.2	+9
Union Carbide	4.9	0.16	–1
Chrysler	36	1.1	+10
Ford	100	4.3	+16
GM	131	5.9	+1
AT&T	65	2.9	–7
IBM	65	5.1	+2

SOURCE: Adapted from *Business Week* (1993a).

TABLE 1.2 Advanced Materials and Processing Program: R&D Funding (then-current-year dollars) by Materials Class

	FY 1991 Expenditure ($M)	FY 1992 Appropriations ($M)	FY 1993 President's Budget ($M)
Polymers	83	83	93
Bio/Biomolecular materials	144	166	187
Composites	185	183	207
Ceramics	137	132	151
Electronic materials	172	162	177
Optical and photonic materials	143	133	139
TOTAL	864	859	954

SOURCE: Federal Coordinating Council for Science, Engineering, and Technology (1992).

prevailing conditions, the committee believes that the discrepancy is more complex.

Is it possible that research and development expenditures now include a broader range of categories? Shifting terminology may also be contributing to the discrepancy. Research has come to include work that is specific to a project and is not expected to reveal new concepts of broad-scale impact. Research and development productivity is increasingly measured in terms of time to market, and time horizons for projects have been progressively reduced (*Business Week*, 1994). This approach could have severe consequences for longer-range projects that provides for the future. Another industrial factor that is not healthy for science is the tendency to reduce central research organizations by assigning research and development people to specific product-line responsibilities. As the research staff is dispersed, the viable nucleus is destroyed. The staff becomes diffuse, and opportunities for essential cross-fertilization are lost. Documentation of these effects is difficult, but the committee is convinced of the reality of the danger to research at large and to polymer research in particular. A recent article (*Business Week*, 1994) comments: "U.S. industry has cut its long-term research by 15% since 1986." The article emphasizes the high rate of growth of Japanese research and development spending, especially that for basic research.

INTERNATIONAL COMPETITIVENESS

Polymers make up a major portion of the chemical industry in the United States. In 1992 the U.S. chemicals group posted exports of $44.0B and imports of $27.7B, for a positive balance of $16.3B (*Chemical & Engineering News*, 1993, p. 71). The corresponding figures for plastics were $10.3B and $4.3B, for a positive balance of $6.0B. (The aircraft industry is the only other major U.S. industry with a significant positive balance of trade.) This trade position is healthy and has been growing, but it may have peaked in 1991. It should be noted that U.S. chemical earnings, profit margins, and return on equity fell over the last 4 years, as markets declined in the recessionary economy (*Chemical & Engineering News*, 1993, p. 48). A detailed interpretation of these statistics is beyond the scope of this report, but it seems clear that the United States is strong in chemicals and in plastics and that this favorable position is in part the result of the U.S. chemical industry's consistent and vigorous commitment to research.

The United States is an acknowledged leader in the polymer markets. The 70 billion pounds of plastic and other polymeric materials produced by the United States annually represent well over half the world's output and more than double the product of the European Community (*Chemical & Engineering News*, 1993, pp. 44, 74, 80). However, of the world's largest chemical companies, all of which make polymers, 4 of the top 5 and 7 of the top 10 are based in foreign countries (*Chemical & Engineering News*, 1993, pp. 50, 75). Further, over half of the 25 companies receiving the largest number of U.S. patents in 1991 were based outside the United States (*Business Week,* 1993b, p. 57). Both patterns indicate strong challenges to U.S. supremacy in the polymer area.

Another area of concern is the U.S. polymer processing industry. Film, fiber, and molded parts constitute three of the largest uses of polymers and the portion of the market that enjoys the greatest value added. However, the equipment required to process the raw polymer into these forms is produced largely in foreign countries, primarily Germany and Japan, and is then purchased by U.S. suppliers and sold under their own label. The Institut für Kunststoffverarbeitung in Aachen, Germany, for example, teaches polymer processing at all levels and develops equipment and processing methods in cooperation with the large German chemical companies. Will nationalism and regional trade advantages combine to give the large chemical companies in such countries a competitive advantage in developing new polymer applications through close association with the equipment manufacturers? Will such developments be protected by patents and thus further erode the U.S. balance of trade?

Two fields of specialty polymers—silicones and fluoroelastomers—developed rapidly after World War II under the leadership of U.S. companies. These high-performance, high-value-added materials have become standards in high-performance equipment used under extreme temperatures or harsh operating conditions. Advances in both fields have been rapid, and U.S. patent positions have been strong, thus effectively keeping foreign companies from gaining market share. But there are strong signs that foreign competition in silicones and fluoroelastomers is gaining, and the steady increase in research and development spending overseas, particularly in Germany and Japan, is serving notice that our lead in this specialty polymer field may be ending. Both Germany and Japan now spend nearly 3 percent of their gross domestic product on research and development, whereas the United States is spending about 2 percent (NSF, 1992).

What is true for silicones and fluoroelastomers is also true for composites. The last four decades have seen continuous growth in composites for a variety of uses. In the early stages, glass fiber was the material of choice because of its good characteristics and low price. As strength-to-weight ratios and stiffness became more important, new fiber reinforcement materials were needed, primarily for uses related to aviation. The ultimate choice for widespread use was carbon fiber. The early work on carbon fibers was done in the United States, but it was not long before foreign companies were also doing research in the field, and today the leading producers are outside the United States, mostly in Japan.

The foregoing competitive problems are merely examples of the difficulties facing the United States as we move toward a global economy. Certainly, every effort must be made to rejuvenate the U.S. industry for polymer processing equipment.

EDUCATION IN POLYMER SCIENCE

The subject of scientific literacy in the United States has generated considerable recent discussion. Within the scientific community, there is a strong sentiment that science education for the nonscientist is woefully inadequate, even at the college level, where few students are exposed to more than introductory courses. The general topic of science education is outside the scope of this report, but concerns about education in polymer science and engineering are closely related. The scope of the problem is large, and modifying the U.S. educational system will require a major cooperative effort over an extended period of time.

Even among most professional scientists, the past level of education in polymer science has been extremely low. This is surprising because it is estimated that about half of practicing chemists and chemical engineers work with polymers at some point in their careers. In the past, most polymer education has occurred as on-the-job training, but the situation is changing. Because of the complexity and interdisciplinary nature of modern polymer science, professionals trained in traditional academic fields cannot be immediately productive upon moving into polymer research or engineering. To fill the need for more professionals with expertise in polymers, the education of future scientists and engineers will have to be modified, and ways will have to be developed for practicing researchers to become aware of rapid progress in macromolecular synthesis, processing, and applications.

Because most students in the United States are educated in traditional areas, it is likely that the key to introducing undergraduates to polymer science will be the natural development of polymer science as an important component of these traditional areas. Exciting and innovative research becomes part of the general curriculum over time. As new applications, strategies for synthesis, theoretical methods, and characterization techniques for polymers become more accessible to faculty in traditional fields, polymer science will emerge as part of the core education in science.

The steady growth of polymers in U.S. industry has resulted in a major increase in polymer science and engineering faculty during the last 40 years. More recently, there has been a significant increase in the number of faculty identified with traditional fields who are conducting research in polymer science. This shift will likely result in the incorporation of concepts of polymer and macromolecular science into a broad range of chemistry and chemical engineering courses.

The number of academic polymer researchers in the United States and Canada can be estimated from survey data published by the Plastics Institute of America (1992), which lists 48 colleges and universities with polymer programs, of which 43 are in the United States. U.S. universities have a total of 548 faculty who are engaged in polymer research. This number is a lower limit because many faculty doing research on polymers are not affiliated with specific polymer science and engineering programs. Even so, a large number of colleges and universities have no faculty with expertise in the polymer area.

A recent survey of the number of polymer faculty appointed from 1985 to 1991 reports that 43 such faculty have been appointed each year, on the average, with no discernible upward trend (Bikales, 1993). It is interesting to note that about half of these appointees have had previous experience in industry. This trend implies that awareness of industrial problems concerned with polymers is growing in many universities. The new faculty have been hired mostly by departments of chemistry, chemical engineering, and materials science, and, in isolated cases, other departments.

Further evidence of the growth of graduate training in polymer fields is shown in Figure 1.4. The number of doctoral degrees granted for polymer research has been steadily rising, with an average increase of about eight per year. Polymer chemistry accounts for about two-thirds of the total number of polymer graduates.

As the introduction of polymer faculty into traditional areas takes place, other actions are assisting the evolution in polymer education. Many of these changes are being fostered by professional societies.

- *Curriculum guidelines.* The inclusion of polymer topics in curricula was endorsed by the 1971 and 1983 guidelines of the American Chemical Society's Committee on Professional Training (ACS-CPT). The 1988 and 1992 guidelines include polymer science and polymer science laboratory as recommended but not as required advanced courses in chemistry. Polymers are strongly recommended as a subject, both for integration into general chemistry curricula and as advanced courses, although the ACS-CPT does not require such incorporation for chemistry curricula to be accredited.
- *Course and textbook development.* A major obstacle to the integration of polymer concepts and examples into existing curricula has been to the lack of polymer-related material in textbooks. A number of new textbooks now provide a chapter on polymers and other macromolecules, most often as topics in physical and general chemistry. POLYED, the joint education committee of the ACS divisions of Polymer Chemistry and Polymeric Materials, has begun a textbook project in which textbook authors and polymer faculty will be encouraged to cooperate. Course development could be facilitated by providing opportunities for interaction between polymer and nonpolymer faculty to help the latter to learn about polymers. These faculty teach most of the fundamental courses and could be encouraged to teach additional courses that have a major polymer content.
- *University faculty development.* How can professors who lack formal training in polymer science enhance their knowledge base so that they can incorporate polymers into their lecture or laboratory courses? Several alternatives are

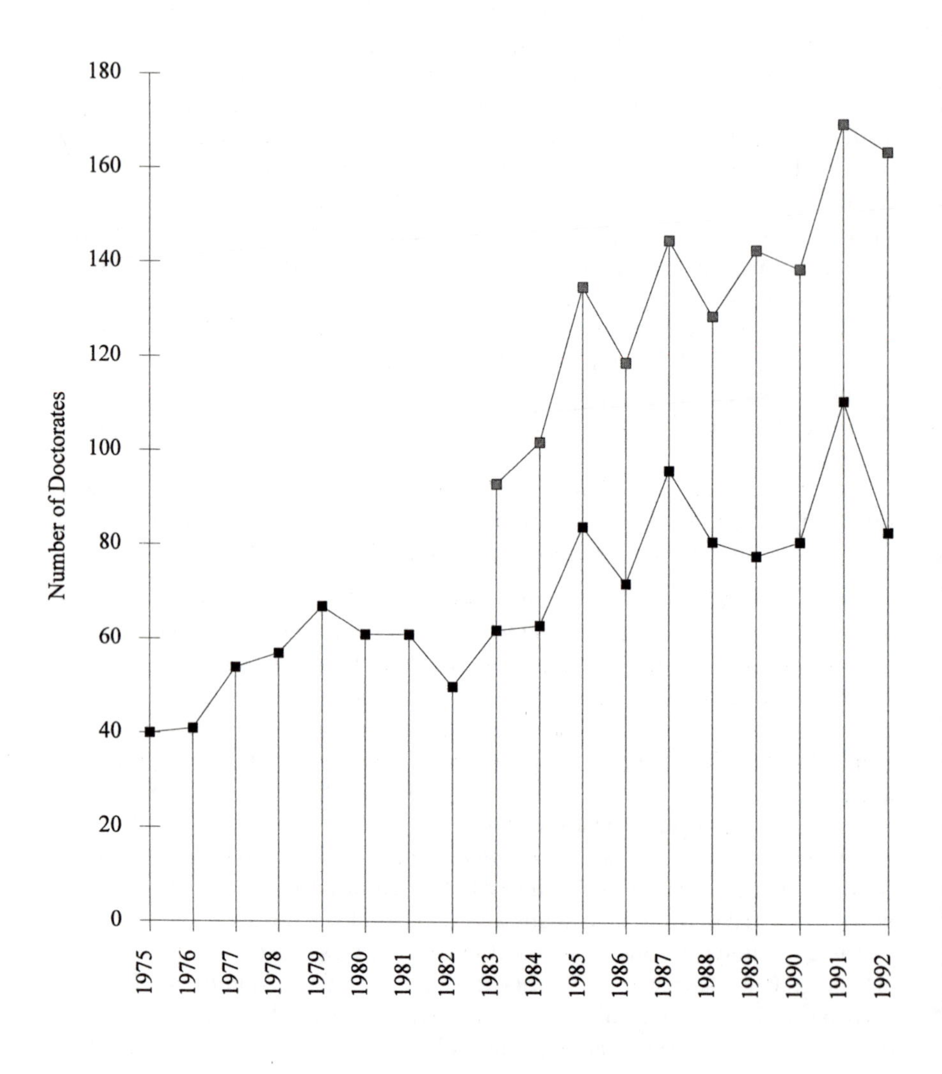

FIGURE 1.4 Doctorates awarded in polymer fields. Polymer chemistry only, solid squares; total polymers (in chemistry, physics, and engineering), shaded squares. SOURCE: Data from Ries and Thurgood (1993).

now available. Two- or three-week summer courses that include laboratory experience have been offered to college-level faculty for several years. In addition, the ACS's POLYED Committee has developed workshops, demonstrations, and new curricular materials.

- *Precollege faculty development.* Several scientific professional organizations have developed programs for primary and secondary teachers (and sometimes their students) that are based on presentations by research scientists from different fields of interest. Some of these activities have a particular focus on polymers. One example is the two-hour polymer workshop and teaching package on polymer chemistry for pre-high school teachers and students that was developed by the National Science Teachers Association. Another program has been developed by the Intersociety Polymer Education (IPED) Task Force, representing the Society of Plastics Engineers and the Polymer Chemistry, Polymeric Materials, and Rubber divisions of the ACS. The IPED program, entitled "Polymer Ambassadors," involves high school teachers as "ambassadors" in North Carolina, Nebraska, Missouri, and Ohio. The ambassadors take part in activities such as visits to elementary and middle school classrooms and workshops for students and teachers.

- *Continuing education.* Many of the chemists and engineers employed by industrial and government organizations work with polymers, even though they received no formal training in polymer science and engineering. To provide needed professional education for this part of the work force, a variety of continuing education courses have been developed. The "Polymer Short Course Catalog" (available from the POLYED Committee of the ACS) lists 113 short courses offered by 28 universities on various aspects of polymer science and engineering. Other technical societies also offer short courses on a variety of topics relevant to polymer education. Video courses are also available. The question remains, however, whether this approach to professional education should serve as a supplement to or as a substitute for training at the postsecondary level.

- *Technical training.* While the United States excels in the production of Ph.D.s in basic science and engineering, the numbers of technicians and scientific assistants trained have not been as great as in other technically advanced countries. Because many of our junior colleges and trade schools have evolved into four-year colleges and universities, there is now a shortage of institutions for educating workers at different levels of training vital to U.S. development and production efforts. The United States does have some technical high schools and two-year colleges, but many of those have fallen behind the needs of industry and do not have the personnel or facilities to adequately train workers needed to participate in polymer development and to carry out sophisticated production operations. By using Ph.D.s to carry out tasks that could be performed by well-trained technicians, the country underutilizes the talents of its highly trained scientists and engineers. Such practices unnecessarily drive up research and development costs and thus jeopardize efficient production of new technologies.

It has been assumed that industry could train its workers for such tasks. However, on-site training may not be adequate to keep pace with rapid increases in technical sophistication. At the same time, the quality of science literacy among graduates from U.S. secondary schools has been compromised. The solution to this problem could involve the development and improvement of technical schools and junior colleges for those who are interested in a technical job, not necessarily a Ph.D. or M.S. position.

- *Industry-university interaction.* Many polymer faculty have previous industrial experience and are aware of industrial issues. Others could benefit from enhanced mechanisms of interaction. Several methods already exist for supporting awareness of industrial research problems in the polymer field: (1) University-industry centers have been established at a number of research universities to investigate research problems of interest to the industrial partners. These centers are particularly valuable for the university partners as long as the research results can be published in a timely fashion. (2) Some industries have hired university faculty as summer researchers. This is a fruitful way to introduce college and university faculty to industrial problems. (3) Opportunities have been provided for industrial researchers to enter academic laboratories for short visits.

Strengthening the competitive position of the United States in polymer science and engineering will require meeting the following future education-related challenges:

- Education of the general public in science, including polymers and other materials. Such an education is essential for meaningful participation in public decisions that involve scientific concepts. The familiarity of polymers makes this area a particularly accessible one in which to introduce scientific concepts.
- Employment of more scientists and engineers with expertise and interest in polymer science and engineering as professors in colleges and universities.
- Facilitation of exchange between academic and industrial laboratories. Polymer science research has been very strong in industrial laboratories.

ENVIRONMENT

The last few years have seen a growing national recognition that Earth's resources are finite and that we share in the responsibility for preserving a reasonable environment for future generations. Unfortunately, no clear, single path of action has been found, because environmental issues often can be solved only locally or on a short-term basis. Although the problems are of massive proportions and too little is currently known, considerable effort is being made, and there are good prospects for progress. Certain aspects of polymer use can be beneficial in this work.

Polymers are organic materials that are usually manufactured from petroleum feedstocks using processes that are environmentally benign or that can be made more environmentally compatible without crippling economic consequences. The polymer industry has successfully adopted many environmentally favorable processes, providing advantages over competing materials as described in the vignette "Paper or Plastic?," which also touches on some of the complicated issues involved in environmental decision making.

Similar complexities are evident in comparisons of polymers with other materials, such as metals and glass. Metal smelting plumes affect the atmosphere for hundreds of miles downstream, and land damage and the disposal of tailings from mining and refining operations are serious problems. Metal refining and glass making require large amounts of energy. Thus, from an environmental standpoint, comparisons of production processes may lead to a preference for using polymers rather than metals or glass. In addition, products fabricated from polymers often weigh less than those made from competing materials. Energy savings favor plastic for bottles over glass, when the energy requirements for production, transportation, and recycling are considered. Substitution of polymers for metals in aircraft and in automobiles results in weight reductions that translate to fuel efficiencies over the life of the vehicle. These efficiencies afford important operating economies and contribute to efforts aimed at decreasing U.S. dependence on petroleum.

Natural materials are not always better than synthetic materials from an environmental point of view. A comparison of fiber production is instructive. The natural fibers cotton and wool make heavy demands on agricultural resources, including land use, fertilizer (from petroleum), and fuel for transportation. Maintenance of clothing made from synthetic fibers requires less hot water. The trade-offs are complicated, but synthetic fibers offer many advantages.

A challenge to the use of all materials is posed by their disposal after use. National concern about the use of landfills has resulted in increased emphasis on recycling programs and a reluctance to use "once-through" products that must be sent to landfills. On a volume basis, polymer waste makes up about 20 percent of current landfill input (by weight, the percentage is lower), and methods must be devised to cope with this large and growing problem. Polymers degrade very slowly in landfills, but under typical landfill conditions even paper does not degrade rapidly enough to match the rate at which it enters the landfill. Although paper and other biodegradable materials degrade much more rapidly under composting conditions, this approach is not generally feasible. Polymers can be made that are degradable at a more rapid, controlled rate, but these materials are not competitive in cost or in other properties, such as durability. Recycling of materials is environmentally attractive but in many cases has not proved to be economically viable. Frequently, products can be made with less expense from

PAPER OR PLASTIC?

While your groceries are being rung up in the supermarket checkout line, the bagger will probably ask you, "Paper or plastic?" Corporations, too, face this decision when packaging their products. In November 1990 the McDonald's Corporation, largely in response to pressure from the public and from environmental groups, made the decision to replace Styrofoam "clamshell" hamburger containers with paperboard boxes. Although immensely symbolic, this decision was not necessarily the right one to minimize the environmental impact of the packaging material. When deciding which product is more environmentally friendly, the environmental consequences of all of the steps of its manufacture, use, disposal, and eventual degradation must be considered, and the results can be surprising. In many ways, the use of paper may be more polluting than the use of plastic, as several studies of their comparative life histories have shown.

Paper is made from wood pulp, a renewable resource, and Styrofoam is made from nonrenewable fossil fuels. But the manufacture of a paper container consumes as much fossil fuel as goes into making a Styrofoam one, so the raw-materials toll of making a paper box includes *all* the hydrocarbon costs of making a plastic clamshell, plus whatever forest degradation is caused by timber harvesting. And, of course, clear-cutting and other unsound logging practices greatly increase this toll by increasing erosion and fouling streams. The manufacturing process uses other resources, too—one study estimates that manufacturing a Styrofoam clamshell uses 30 percent less energy, and generates 46 percent less air pollution and 42 percent less water pollution, than does manufacturing a paperboard box. And since paper products generally weigh more than their plastic counterparts, more fuel is needed to ship an equivalent number of paperboard boxes to the restaurant, and to pick up the used ones later. There is also the performance issue—Styrofoam is a better insulator than paperboard, so the food stays hotter longer. Food in paperboard containers will need to be reheated sooner, consuming more energy.

Once the product has been used, there is the problem of disposal. Paper products that have been treated to repel grease have very low recycling value, as the coating agents interfere with the recycling process. And it has recently been discovered that paper in landfills degrades much more slowly than had been thought—newspapers thrown out in the 1950s were still readable when unearthed some 30 years later. Plastic, however, can easily be recycled into new products or used as a fuel that burns cleaner than coal or oil, thus recovering the fossil-fuel value of the petroleum in the plastic.

Much of the objection to Styrofoam stems from the belief that its manufacture requires the use of chlorofluorocarbons (CFCs) that harm the ozone layer. This is no longer true, as industry stopped using CFCs in Styrofoam manufacture many years ago.

In summary, replacing Styrofoam with paperboard, which on the surface seems a simple, environmentally friendly decision, involves a complex set of environmental trade-offs. So, paper or plastic? Making the correct decisions requires a comparative analysis of the environmental impacts of the life histories of the alternative products. Your decision at the supermarket is similarly complex.

raw materials. In part, this observation may reflect our imperfect accounting systems, in which the producer bears no responsibility for the cost of disposal. Manufacturing firms are now moving to design and engineer new products with an eye to the need for recycling at the end of product life. The European automobile industry is a leader in this area, and U.S. automobile manufacturers have formed the Polymer Recycling Consortium to coordinate similiar work.

Recycling often involves the production of lower-grade products from recycled feedstocks. This degradation in quality has several sources: the recycled material may have been degraded to some extent in the first use, it can be further degraded by reprocessing, and the mixtures of polymers that are frequently formed during recycling often have seriously diminished physical properties and appearance. Success in recycling depends very much on the chemistry of the polymer involved and on the purity of the scrap. One outstanding example is polyester soda bottles, whose recycling has been aided by "bottle bills" that encourage their return in relatively uncontaminated form. Labeling of plastics at the time of manufacture has been initiated to facilitate separation, and methods of automated recognition and separation are now being sought. Factory scrap is often fed back into the feed supply to achieve an economic advantage with no sacrifice in quality. Despite considerable effort and progress, however, recycling of plastics generally poses a major unsolved problem.

Incineration is another option for disposal that can be attractive in certain circumstances. Polyolefins, for example, have high fuel value and can be burned cleanly and be disposed of with carbon dioxide and water as the only products. But other polymers pose more difficult problems for incineration. Combustion of poly(vinyl chloride), for example, produces corrosive and toxic smoke that requires complex and expensive scrubbing, followed by subsequent disposal of the scrubbings. Also, the fuel content of this polymer is lower, and net energy consumption is required. Other polymers present different scrubbing problems, and mixtures of polymers complicate incinerator operating strategies. Some progress is being made, but the advances are evolutionary. Nevertheless, until improved strategies for recycling have been implemented, incineration offers many advantages for disposal of plastics.

In recent years, there has been an active effort to invent new polymers that will have the needed properties during use but will decompose by some specific process, such as biodegradation or photodegradation, when disposed of. Some progress has been reported, but the requirements for conflicting properties means that this approach will not be applicable to every situation. Is it reasonable, for example, to hope for a high-strength fishing net that falls apart only after it has been lost or discarded? On the other hand, there may be special situations in which degradable polymers will be valuable, a good example being the use of plastic "six-pack rings" for packaging soft drink and beer cans. A plastic that degrades when exposed to sunlight for just a few weeks presents a minimal hazard to the environment. Another possible example is that of polymers based in part on starch. These materials are often water sensitive and may offer a way to produce biodegradable polymers. Such materials would be more susceptible to composting than most synthetic polymers, but their properties appear to be limited. In principle, polymers that degrade at controlled rates are possible, but the logistics of their use can be a problem. In addition, if degradable polymers are mixed with other polymers, the resulting mixture will be less satisfactory for recycling.

HOUSING AND CONSTRUCTION

Polymers have made important inroads into the construction of housing and other buildings, despite the difficult competition presented by traditional materials—such as wood, brick, stone, concrete, glass, asphalt, plaster, slate, iron, and copper—that are economical and are entrenched in terms of aesthetics, artisans' skills, and existing U.S. building codes. The market has been penetrated by synthetic polymer products, often used in combination with traditional materials, across a broad range of applications. Demonstration houses already exist in Massachusetts and New Jersey that showcase an impressive array of uses, as pointed out in the vignette "Plastic Houses."

As the acceptance of factory-built (prefabricated) structures grows, economies of scale will arise that will strongly favor the use of polymeric materials in many areas. Complicated shapes can be fabricated economically if the costly tooling can be depreciated over large numbers of copies. Synthetics offer greater uniformity than natural materials, a factor that will grow in importance as factory production increases. Further, as synthetics continue to improve, it is likely that polymers, composites, and combinations will become superior in those properties that are valued by the consumer.

Polymers have already become dominant in fabrics for household furnishings. Carpeting, draperies, and upholstery are now usually made of synthetic polymers because of their superior properties. Substitutions are also evident in appliances such as refrigerators, coffee makers, and mixers.

Progress in the use of polymeric materials in housing will probably continue to be incremental, as each application is tested on the basis of its merits with existing materials of choice. Although societal inertia is particularly large in the construction trades and housing, the movement to increased use of polymer-based materials is unmistakable.

ENERGY AND TRANSPORTATION

The United States consumes energy out of proportion to the size of its population, and a large fraction of this energy is consumed by our extensive transportation system. While rail is still significant for shipping freight, the highway network is the defining transportation mode for the twentieth century. Automo-

PLASTIC HOUSES

The new house you buy 20 years from now could be made almost entirely of plastic. A demonstration house, built to show the feasibility of such a scheme, and to test new materials and construction methods, already exists. And the house is environmentally friendly—high-strength plastics, such as polycarbonate and acrylonitrile butadiene styrene (ABS), reclaimed from junked cars and discarded packages, were melted down and reused to construct everything from the outside walls and shingle roof to the toilet bowls. Even the concrete in the basement is 60 percent recycled plastic.

On the interior, standard-sized plastic polyphenylene oxide wall panels are held to the studs with Velcro for easy access to plumbing, electrical, heating, and other utilities. Light switches and electrical outlets can be formed right into the panels. Heater ducts, molded into the panels and abutting a metallized plastic film just inside the wall's surface, turn the whole wall into a radiator. And the modular system allows panels incorporating built-in cabinets, desks, shelves, and so on to be added whenever, and wherever, they are needed.

The kitchen and bathrooms feature lightweight, superdurable countertops made of concrete-plastic, an alloy of poly(ethylene terephthalate) and poly(butylene terephthalate) aggregate, finished with a silicone sealer so stains and spills wipe away. Standardized sizes, modular units, and snap-together plumbing make remodeling the kitchen as easy as rearranging the furniture. The subfloors throughout the house are blown foam tiles, made of poly(phenylene oxide), imprinted with a pattern of raised squares like a waffle iron. The channels between the squares accommodate the plumbing, and the squares themselves are like the knobs on a Lego® block, allowing items like bathtubs and cabinets to be snapped into place. The windows are actually sheets of a transparent liquid crystal sandwiched between panes of polycarbonate plastic 250 times stronger than glass. The liquid crystal turns from clear to cloudy at the flick of a switch, making curtains unnecessary for privacy. And a tinted film of a polyester liquid crystal on the windows prevents excessive heat buildup indoors and keeps carpets and furniture from fading.

The parts for such houses would be manufactured in two plants. Large, centralized plants would receive the raw, recyclable plastic, grind it up, and form it into new roof, wall, floor, and foundation panel stock on a continuous-feed line. The stock would be cut into 8 by 40 foot panels and shipped to local finishing factories, where the panels would be customized according to the architects' specifications for the houses being built in the area. Bundles of finished panels, cut to size and with doors and windows already installed, would be shipped to the construction site, ready for quick assembly. The ease and economy with which such houses could be constructed may give new meaning to the term "affordable housing."

biles for personal transportation and, increasingly, trucks for freight have become ever more important. Air transport dominates business travel, much long-distance personal travel, and priority parcel and freight delivery. All of these forms of transportation depend heavily on oil as the basis for fuel.

Polymers play a significant role in energy and transportation, primarily through the savings in weight that are possible as polymers replace steel and

POLYMERS IN LUBRICANTS

The multigrade oil in your car's engine is not just the gooey remains of some long-dead dinosaur sucked from the sands of Saudi Arabia. It also contains a polymer additive, usually about 1 percent in concentration, that modifies the viscosity-temperature relationship of the base oil. This esoteric-sounding bit of engineering is critical to your car's health. As those who have ever heated maple syrup for their pancakes know, the hotter a viscous liquid is, the runnier it gets. Unfortunately, what your car needs is an oil whose viscosity is essentially constant—low enough to be pumped through the engine, yet high enough to lubricate and protect the moving parts—no matter what the temperature is. But an unmodified light oil that is runny enough to pump when the engine is cold will be much too thin to lubricate it once it warms up. (Typical engine operating temperatures hover around 150°C.) Similarly, a heavy oil that lubricates well at 150°C is simply too viscous, like molasses in January, to work in a cold engine.

Fortunately, polymers have come to the rescue. Even low concentrations of a polymer in a solvent—including oil—can increase its viscosity quite a bit. It is possible to make a solution whose viscosity does not change very much with temperature. The mechanism for this ability of polymer additives to provide us with multigrade motor oils is complex; in fact, quite a few mechanisms may be operating at the same time. Some of these polymers can simultaneously be used as dispersants for water and the sulfuric acid that gets into the motor oil from burning sulfur-containing fuels. These are usually graft copolymers of an oil-soluble polymer with a water-soluble polymer.

These principles say nothing about how much polymer to use, or how long its molecules should be, or what they should be made of. In practice, the chemically active environment of a hot engine, which is sufficient to break down many polymers, and the oil solubility and cost of manufacturing those polymers that can withstand a hot engine narrow the choices considerably, but there are more subtle considerations. The shearing force felt by a polymer molecule trapped between a static engine block and a plunging piston may be enough to tear the molecule apart. As the polymer gets longer, its susceptibility to shear increases, putting an upper limit on its size. However, smaller polymers have less thickening power. It takes more of a smaller polymer to produce the same increase in viscosity, and so some balance must be struck between the polymer's effect on viscosity and its length.

The polymer is not the only additive, either. Other chemicals, such as dispersants, antioxidants, and wax crystal modifiers, are also needed. These additives, unfortunately, interact with the polymer to diminish its effectiveness in cold engines. For this reason, and because it is easier and cheaper to add one thing to an oil instead of several things, there is considerable interest in creating multifunctional polymers that could take over the roles of the other additives, or at least not interact with them. There is plenty of exploration left in the oil business for creative polymer chemists.

aluminum in vehicle structures. Polymer composites are 5 to 7 times less dense than steel and 2 to 3 times less dense than aluminum. The advantages of weight reduction will continue to be most critical for automobiles and aircraft. Automobile weight is substantially in the vehicle itself, while for trucks and freight trains the weight is largely in the payload. Impressive fuel-saving reductions in the weight of automobiles over the last two decades are attributable to the substitution of polymers and polymer composites as well as high-strength steel alloys, and much of the credit for improved fuel efficiency can be traced to the introduction of these advanced materials. Present-generation automobiles are about 10 percent polymer by weight, and the potential for further substitution is substantial.

The potential is greater yet for aircraft. For each pound saved in the weight of the airframe, the payload can be increased by a pound, and over the life of the aircraft the payback is large. Polymeric materials have made important inroads in general aviation. The airframe of the Beech Starship, for example, is almost entirely polymer and polymer composite, demonstrating the viability of polymers for all nonengine aircraft parts. Less progress has been made in commercial and military aircraft, which follow more conservative design philosophies. But the potential is evident, and extensive substitution can be predicted with confidence.

Further progress in substituting polymers in the transportation area will be based on the high strength and durability of polymers and polymer composites. Further progress in manufacturing processes, repair procedures, and failure mode control will enhance the use of polymers. There is no fundamental reason that polymeric materials cannot be introduced in all areas other than those that expose the materials to very high temperatures, such as engine parts and some aircraft surfaces. One of the best-established uses of polymers is the use of polymeric elastomers in tires, a critical application that affects safety.

Polymers are widely used in the form of foamed insulation, especially in buildings and refrigerators. Although there are concerns about fire and smoke hazards, this use will probably increase. The use of chlorofluorocarbons (CFCs) as blowing agents and foam cell fillers has largely been phased out to reduce ozone layer depletion. Polymers are also used as components in other energy devices and equipment, such as solar systems, fuel cells, batteries, and even reactors. Applications of polymers are discussed in the vignettes "Polymers in Lubricants" and "Polymers for Oil Recovery."

Polymer science and engineering will continue to have a significant impact on energy conservation and on transportation. There is further need for development of better materials having unique properties for applications in these areas. Many will be high-value-added applications, thereby justifying a significant research effort and the use of more expensive materials.

NATIONAL DEFENSE

The military uses polymers widely in diverse applications. These include clothing such as bulletproof vests and helmets, now made exclusively of such polymers as Kevlar® and Spectra®. Structures such as tents, huts, and bridges are made of polymers, as are aircraft, ground vehicles, and naval structures.

As in civilian aircraft, the use of polymers in military aircraft would reduce their weight, thereby conferring the key competitive advantage of longer range and increased payload (for armaments, ammunition, and electronic equipment).

POLYMERS FOR OIL RECOVERY

We have all seen pictures of a "gusher," the spurt of black oil that sometimes erupts from an oil well when it first penetrates the oil-bearing stratum in the rock below. This oil is under great pressure and will continue to flow up the well pipe on its own accord until the pressure equalizes. Once this happens, the oil can still be pumped out, using enormous pumps that look like bobbing birds. Eventually, no more oil can be pumped, but at that point there is still plenty of oil down in the porous rock stratum.

Once the pumps have run dry, the oil companies use "secondary" recovery processes—pumping water into the well to displace the oil upward, for example. But water injection does not recover most of the oil that is still down below. The water, being thin and runny, does not push the thick, viscous oil very well. Instead, the water flows around and past the oil, which stubbornly clings to the rock. On average worldwide, only about 25 percent of the oil in a stratum is recoverable by primary and secondary means, although in some cases the yield can be as high as 50 percent.

Sometime around 1960, somebody suggested that adding a water-soluble polymer to the water to increase its viscosity might help it push the oil along ahead of it. Sure enough, the thickened water can no longer flow around the oil, and another 10 to 20 percent of the oil in the rock can be recovered. A partially hydrolyzed form of the polymer polyacrylamide is usually used, although xanthan gum—a biopolymer more commonly found as an additive in foods—is also employed. Other polymers are also being developed whose properties are especially designed for enhanced oil recovery. As world oil reserves are depleted, these polymers will become more and more important in the oil companies' quest to get the most from every well.

Extension of range and payload will occur for each pound of polymer-based material that is substituted for a greater weight of aluminum or titanium. Failure of the airframe material is a significant concern, but polymer matrix and polymer- or carbon-reinforced composites offer a wide range of behavior that could provide an optimal mix of properties. The high skin temperatures that are generated at supersonic speeds present another challenge, but there are many new high-temperature matrices that, used with carbon fibers, could provide novel solutions.

The leverage obtained by using polymers is greater for ground-based military vehicles than for aircraft, because polymer-based composites are being substituted for steel, which is about 5 to 7 times as dense as the composites. Military design philosophy has been strongly influenced by the need for protective armor plates. But the advent of armor-piercing artillery shells suggests shifting to a protective strategy based on light weight for range and speed, coupled with deflection of antitank and antivehicle missiles by features of vehicle architecture, such as low-angle surfaces.

Military marine structures and vehicles are still built largely of steel, which demands heroic defenses against corrosion and very heavy (nonplaning) hull structures. Polymer-based substitution could be highly advantageous in many respects. It is no accident that civilian boats are now manufactured almost entirely from polymer composites. The potential advantages are attractive for high-speed craft such as hydrofoils, particularly since corrosion problems would be eliminated. As the effectiveness of armor-piercing projectiles reduces the importance of bulk armor, polymer-based materials should come to the fore. Unmanned vehicles will become cheaper and more effective. The performance of shallow-draft vehicles for river and beach landing operations will be enhanced by light weight. Transport of marine (as well as land) vehicles will become faster and easier with lighter weight. Polymer-based hull coatings that afford a low-friction skin for marine vehicles could provide a military breakthrough.

Achievement of a low radar, sonar, optical, or source signature is a critical military objective. Aircraft need a low radar reflective surface. Ships and particularly submarines need a low sonar reflective surface. Land-based vehicles also need these features. Polymer-based materials can be uniquely adapted and tailored to achieve the stealth objective. This is an active but still emerging field.

REFERENCES

Bikales, N.M. 1993. "Herman Mark's Children." *Journal of Polymer Science: Polymer Symposia* 75:143-150.

Business Week. 1993a. "R&D Scoreboard: In the Labs, the Fight to Spend Less, Get More." June 28, pp. 102-127.

Business Week. 1993b. "The Global Patent Race Picks Up Speed." August 9, pp. 57-58.

Business Week. 1994. "Who Says Science Has to Pay Off Fast? Japan Is Pursuing Long-Term Projects—Just As the West Backs Off." March 21, pp. 110-111.

Chemical & Engineering News. 1993. "Facts & Figures for the Chemical Industry." Vol. 71, No. 26, June 28, pp. 38-83.

Federal Coordinating Council for Science, Engineering, and Technology (FCCSET). 1992. *Advanced Materials and Processing: The Fiscal Year 1993 Program.* A report by the Office of Science and Technology Policy, FCCSET Committee on Industry and Technology. (Available from Committee on Industry and Technology/COMAT, c/o National Institute of Standards and Technology, Rm. B309, Materials Building, Gaithersburg, MD 20899.)

National Science Board (NSB). 1992. *The Competitive Strength of U.S. Industrial Science and Technology: Strategic Issues.* A report of the NSB Committee on Industrial Support for R&D. NSB 92-138. Washington, D.C.: U.S. Government Printing Office.

National Science Board (NSB). 1993. *Science & Engineering Indicators 1993.* NSB 93-1. Washington, D.C.: U.S. Government Printing Office.

National Science Foundation (NSF). 1992. *National Patterns of R&D Resources: 1992.* NSF 92-330. Washington, D.C.: U.S. Government Printing Office.

Plastics Institute of America. 1992. *1991-1992 Directory of Polymer Science and Engineering Programs.* Plastics Institute of America, Fairfield, N.J. (Available from Stevens Institute of Technology, Fairfield, N.J.)

Ries, P., and D.H. Thurgood. 1993. *Summary Report 1992—Doctorate Recipients from United States Universities.* Washington, D.C.: National Academy Press. (The Survey of Earned Doctorates is conducted for the following agencies of the U.S. government: National Science Foundation, U.S. Department of Education, National Institutes of Health, National Endowment for the Humanities, and U.S. Department of Agriculture.)

2

Advanced Technology Applications

New classes of polymeric materials with unique applications are being introduced. In many cases, the properties and their usage were discovered only recently. This chapter covers two areas: (1) health, medicine, and biotechnology, a rapidly developing domain based largely on known materials but moving to designed and engineered polymers, and (2) information and communications, an emerging field for polymers significantly based on their electronic properties. These two areas are attracting a great deal of attention, particularly among researchers who are not traditional specialists in polymer science. The growing importance of these fields makes the interdisciplinary aspect of polymer research abundantly clear.

HEALTH, MEDICINE, AND BIOTECHNOLOGY

Polymers play a major role in all aspects of biological processes. In fact, it is legitimate to proclaim that polymers are the molecular basis of life. The genetically inherited information required for the growth and health of living systems is encoded in the macromolecule deoxyribonucleic acid (DNA), the backbone of which forms the famous double helix. The molecular genetic code uses only four purine and pyrimidine bases to dictate the structure of the proteins that make up so much of living systems. DNA directs the assembly of about 20 amino acids in complex sequences that become the proteins. These proteins are polypeptide polymers that differ from one another only in the sequence of their constituent amino acids. All enzymes, which control the reaction rates in biological systems, are proteins. Collagen proteins form fibers and connective tissue found in tendons, cartilage, blood vessels, skin, and bone. Elastin, an elastic substance found in ligaments and in the walls of blood vessels, is also a protein. Other polymers such as polysaccharides are also important. They make up chains of sugar units present as a major constituent in all connective tissue. Ribonucleic acid (RNA) molecules also carry information and can serve protein-like functions. Thus informational, chemical, mechanical, and other properties of living systems find their origin in the molecular structure of their component polymers.

Medicine, as a biological science, therefore must be dependent on the nature of polymers. Bandages and dressings are dominated by polymers in modern practice. Molds and impressions of teeth, dentures and denture bases, adhesives, and fillings are polymer based. Sutures, which were made of cat gut for over 2,000 years, are now made of synthetic polymers. Hard and soft lenses required after cataract surgery, artificial corneas, and other ocular materials are all polymers. Orthopedic implants, artificial organs, heart valves, vascular grafts, hernia mesh, and artificial arms, legs, hands, and feet all depend critically on polymeric materials. Similarly, catheters, syringes, diapers, blood bags, and many other trappings of modern medicine depend heavily on polymeric materials. Most of these items arrive in sterile form, packaged in polymers.

Significant quantities of polymers are used in medical devices, consumable medical products, and the packaging for medical products. The most common products are devices such as catheters and intravenous lines, nearly 100 million of which are used annually in the United States. Because medical products use functional rather than structural polymers and their value is not related to the volume they occupy, medical products should be quantified on the basis of number of functional units rather than in terms of pounds of polymers.

Polymers are natural allies of medicine because living tissue is composed substantially of polymers. As our understanding of the processes of life advances, and our ability to tailor synthetic polymer structures to specific chores matures, the power of medicine will grow dramatically. The opportunities for collaborative programs involving materials scientists and medical researchers and practitioners are unlimited. Few, if any, areas of research offer more obvious benefits to society.

Medical devices generally entail intimate contact with living tissue. Organisms are extremely sensitive to the presence of foreign substances and are aggressive in repelling an invading object or agent. To date, empirical means have provided considerable progress in finding materials that are less offensive to living organisms. Polyesters, polyamides, polyethylene, polycarbonate, polyurethanes, silicones, fluorocarbons, and other familiar polymers have been employed successfully in medical applications. Establishing the factors controlling the biocompatibility of these materials is a difficult process that has been only partially defined. Materials experimentation in medical applications has always called for courage as well as technical know-how, but in this litigious era the problems are amplified. Even so, progress continues on a broad front.

Advances in any technology depend critically on research. Currently, there is a paucity of fundamental information on interactions between synthetic materials and the biological medium, partly because of the complex mechanisms involved and the fact that the human body is a most hostile environment. Hence, there is a need for research that will generate the fundamental information necessary to design materials that will be compatible with human tissue and perform the required functions. In particular, there is a need for better understanding of surface interactions and relationships between physical properties of a polymer and biological events, such as clot formation in blood.

A persuasive argument can be made that biomaterials development is poised for a revolutionary change. For quite some time, an important objective of biomaterials research has been a search for "inert" materials that elicit minimal tissue response. But nothing about organisms is static. As the processes of cellular signaling and differentiation become more thoroughly understood, it is likely that new polymers will be engineered to manipulate these processes in positive, productive ways. The emerging science of tissue engineering, for example, will depend directly on the development of new biologically active polymeric matrices to guide the controlled generation (or regeneration) of skin, cartilage, liver, and neuronal tissue. Enzymes, semisynthetic enzymes, and genetic engineering provide a revolutionary opportunity in the production of novel materials for medical uses. The challenges are great; the rewards are greater. The potential economic and societal impact of polymers designed for use in health care, biotechnology, and agriculture is enormous. Almost everyone is already in contact with biomaterials. Hence, polymers are positioned to play a vital role in improving the quality of life, enhancing longevity, and reducing the cost of health care.

Polymers in Health Applications

Implants and Medical Devices

Development of medical implants has been limited by many factors. The synthetic nondegradable materials needed in such products as orthopedic joints, heart valves, vascular prostheses, heart pacemakers, neurostimulators, and ophthalmic and cochlear implants must meet many technical requirements, including being stable and biocompatible in the host environment for moderate to long lifetimes. However, the fact that most of the polymers currently used in implants were not initially designed for medical use means that those polymers may not meet such requirements. This carries with it inherent risks, such as those dramatically brought to light in the course of recent litigation concerning silicone breast implants. Also, toxic breakdown products have been reported for certain polyurethanes under consideration for an artificial heart pump design. Development of new techniques for screening and testing the biological response of candidate materials is clearly a priority matter. By and large, however, empirical testing has found material implants to be remarkably successful.

Each implant application calls for a specific set of properties. A key property required in implants exposed to blood is nonthrombogenicity; that is, the material must not cause clotting. Polymers being considered for vascular prostheses include poly(ethylene terephthalate) fibers, expanded polytetrafluoroethylene foams, segmented porous polyurethanes, and microporous silicone rubber. Surface treatments include hydrophilic coatings, seeded endothelial cells, immobilized heparin (an anticoagulent), and a garlic extract. The diversity of candidates is impressive. Although none of these materials are completely satisfactory, good blood flow has been maintained for many years in some cases.

Polymers also play a major role in devices used to oxygenate blood. They must operate without blood damage. Silicone rubber and polypropylene have been used successfully in both solid and microporous forms. These materials, in microporous form, are widely used during cardiopulmonary bypass surgery, where blood exposure is relatively short term. For long-term exposure, solid membranes are used. Again, surface treatments, such as immobilized albumin, are providing promising results.

Synthetic, biomimetic phospholipid membranes are under development as coatings that render surfaces compatible with blood. Inclusion of the phosphorylcholine headgroup is thought to be a most promising approach, and it has been employed on poly(vinyl chloride), polyethylene, polypropylene, and other polymers. The phosphorylcholine group can also be added as a plasticizer in polyurethanes and other polymers.

Artificial kidney machines employ polymeric hollow fibers to purify blood by hemodialysis. Cellophane (regenerated cellulose) was introduced early on, and Cuprophan, a form of regenerated cellulose that has been strengthened by cuprammonium solution treatment, remains the material of choice, although many other polymers have been tried. Many factors are involved, including treatment of the dialyzer for reuse and avoidance of removal of desirable factors from the blood. It seems likely that synthetic polymers will eventually come into use, although to date they do not have the proper combination of properties.

Dental materials are dominated by polymers to an increasing extent. Impression materials are made of silicone and polysulfide elastomers that cure rapidly in the mouth and maintain their shape. Denture bases are made from polymers based on poly(methyl methacrylate) (PMMA) that are cross-linked through a free radical process. Fillings that match the teeth in appearance are composed of highly filled difunctional methacrylates that are cured by exposure to blue light. Silane-coated ceramic fillers provide the visual match and the hardness and durability required. The use of photocuring relieves the dentist of the need to work within the limited time allowed by amalgam fillings. The composite has been engineered to minimize contraction during cure, an extremely important aspect of any filling material. Polymers also play a central role in dental adhesives. Further advances in dental materials can be expected as polymer systems are designed and engineered to satisfy the complex needs of the area. The vignette "Dental Composites" further describes the role of polymers in dentistry.

The general field of load-bearing implants involves metals, ceramics, and polymers, and the field has advanced rapidly in recent years. Hip replacements

DENTAL COMPOSITES

When you visit your dentist for a new crown or a set of dentures, you may go home with a mouthful of plastic. Traditionally, crowns for teeth in the back of the mouth, where strength is more important than appearance, have been cast from alloys of mercury with silver or gold. And dentures have been made with porcelain pearly whites rooted in a pink base of an acrylic polymer—a lifelike combination that is rugged enough to chew ice cubes, while the firm plastic base distributes the stresses gently. Fitting these crowns and dentures is a time-consuming process, because they cannot be made to order in your mouth. But some dental work *has* to be custom-made on the spot, and that is where nothing but a polymer will do.

When a dentist is trying to repair a chipped tooth, say, in the front of the mouth, it is essential that the replacement material not only look like a tooth, but also be capable of being molded in the mouth to what is left of the original tooth. The material must be strong enough to chew with and should seal the tooth's interior from decay-causing bacteria and from hot, cold, or other potentially painful foods. For perhaps 100 years, the material used for this purpose was "silicate cement," a composite of glass particles held in an acidic gel matrix. This material worked just fine when brand new but would gradually erode over the years.

In the 1950s the technology was developed to allow a polymeric prosthesis to be cast directly onto a properly prepared tooth. These plastics, properly colored, looked just like real teeth and did not decay. Unfortunately, they did have other problems. Methyl methacrylate, for example, a high-strength polymer used to make Plexiglas, would expand slightly in a mouthful of hot soup and shrink when exposed to ice cream, so that the filling would eventually leak, allowing the tooth to decay underneath it. And the polymerization reaction itself liberated a lot of heat—enough to burn a few unsuspecting patients' tongues! The thermal changes were overcome by incorporating high concentrations of glasslike filler particles into the polymer, but these materials were not strong enough to last long because the glass and the polymer did not stick to each other very well. The essential step in developing successful dental composites was finding a suitably strong polymer that also adheres to glass.

Today's composites are based on a dimethacrylate monomer that has side groups dangling off its backbone that are adsorbed onto the surface of the glass particles. And the particles themselves are coated with a coupling agent, such as a silane, that promotes binding. Other "wetting agents" encourage the polymer to seal to the tooth's enamel and the dentin below it. The composite's precursor, a kind of putty that can be applied to the tooth with a trowel-like dental probe, also contains inhibitors that prevent the material from setting prematurely. Once the composite has been properly sculpted, an ultraviolet light is used to initiate the polymerization. Within minutes, the material gels, and after a few more minutes of reaction time the hardening is complete and the patient can go home.

have become common, and satisfactory performance is experienced for decades. A PMMA material system similar to that employed in the fabrication of denture bases is used to bind metal hip replacements to the femur. Ultrahigh-molecular-weight polyethylene is used as the hip cup material. Even the metal alloys of the bone replacement are gradually beginning to be replaced by fiber-reinforced composites. Knee replacements have also become much more common and successful in recent years. This general area has made considerable progress based on increased understanding of bone growth processes that aid bonding to the prostheses. This is a huge field that is advancing rapidly as new and superior materials are introduced.

In the eye, materials are not brought into contact with blood. Contact lenses are external to the body, but the materials are maintained in intimate contact with tissue. Glass was used for many years, giving way to PMMA beginning in the 1940s. PMMA is a hard, glassy polymer that is compatible with the surface of the cornea. Soft lenses made of poly(2-hydroxy ethyl methacrylate) or simply poly-HEMA hydrogel have become popular more recently, in part because of their high oxygen permeability. Poly-HEMA will take up water to a high degree and become soft and flexible. Soft contact lenses contain about 70 percent water. The polymers currently in use are copolymers of vinyl pyrrolidone with poly-HEMA or PMMA.

Replacement lenses provided following cataract surgery are made of similar polymers. The clouded lens is removed and replaced by a hard lens (PMMA) or a soft hydrogel lens. The hydrogel lens may be inserted through a smaller incision, but it has a smaller refractive index than that of PMMA, requiring a greater thickness. An alternative procedure involving injection of a prepolymer liquid into the lens capsule and polymerization in place has been studied. The introduction of new polymer materials continues to make cataract surgery and recovery of sight safer, less distressing, and more effective.

Diagnostics

Polymers are used in diagnostics either as reagents or as enhancers. Polymeric materials can enhance performance of test materials. They are used as solid supports to bind the material being tested specifically for isolation and detection. In other uses, they serve a "reporter" role. The bioreagents are generally incorporated into the system through direct attachment via either copolymerization or cross-linking. The resulting aggregate has multiple copies of the reactive signal and thus can influence accuracy, testing time, and automation. The extent of incorporation will affect diffusivity and exchange rates of solutes, nonspecific binding, and overall binding capacity. The development of automated clinical analyzer (ACA) systems, for example, relied on the heat sealing and good optical properties of the ionomeric polymer called Surlyn®. Hence, the availability of relatively inexpensive polymers will positively influence the development of disposable diagnostic test kits. Polystyrene, nylon acrylamide, dextrans, and agarose have all been used for attachment of antibodies and antigens. In all of these uses, nonspecific binding has to be minimized because it limits sensitivity and makes interpretation of test results very difficult. Therefore, the need to understand interfacial biointeractions will continue to be paramount.

New materials, such as block copolymers containing polypeptides and segmented poly(ether urethanes), have been shown to have specific affinity for proteins. These hybrid materials may prove to be one of the best ways to incorporate both function and structure into the same molecule. For example, it may be possible to incorporate a specific cell-binding segment of a protein into a synthetic polymer, with the latter providing the scaffold and processing capability. Thus, one can tailor polymers to specific biomolecular and diagnostic functions.

Controlled Drug Release

Interest in drug delivery research is increasing for a number of reasons: the need for systems to deliver novel, genetically engineered pharmaceuticals, the need to target delivery of anticancer drugs to specific tumors, the need to develop patentable sustained delivery systems, and the need to increase patient compliance. Polymers are essential for all the new delivery systems, including transdermal patches, microspheres, pumps, aerosols, ocular implants, and contraceptive implants. The major disease areas that are expected to benefit from development of new delivery systems include chronic degenerative diseases, such as central nervous system disorders associated with aging, cancer, cardiovascular and respiratory diseases, chemical imbalances, and cellular dysfunction. Delivery to difficult-to-reach areas such as the brain is desirable, and progress is being made in the area through the use of polyanhydrides, as is discussed in the vignette "Implanted Polymers for Drug Delivery." Success in this area will be rewarded with improved quality of life and longevity.

Several drug release technologies have become clinically and commercially important. They can be classified into various categories by their mechanism of release: (1) diffusion-controlled systems, where the drug is released by solution-diffusion through a polymeric membrane or embedded into a polymeric matrix where the matrix controls the rate of delivery from the system; (2) erosion-controlled systems, where the drug release is activated by dissolution of the polymer, disintegration of the polymer, or chemical/biodegradation; or (3) osmotically controlled systems, where the contents are released by the rate of osmotic absorption of water from the environment, thereby displacing the drug from the reservoir. Any of these mechanisms can be employed to develop controlled-release delivery systems for oral, transdermal, implant insert, or intravenous administration, although some mechanisms are superior to others for cer-

IMPLANTED POLYMERS FOR DRUG DELIVERY

We have all heard that biodegradable polymers are good for the environment. But they may be good for cancer patients, too. Efforts are now under way to design polymer implants that will slowly degrade inside the human body, releasing cancer-fighting drugs in the process.

Such an implant would need several specific properties. It would have to degrade slowly, from its outside surface inward, so that a drug contained throughout the implant would be released in a controlled fashion over time. The polymer as a whole should repel water, protecting the drug within it—as well as the interior of the implant itself—from dissolving prematurely. But the links between the monomers—the building blocks that make up the polymer—should be water-sensitive so that they will slowly fall apart. Anhydride linkages—formed when two carboxylic-acid-containing molecules join together into a single molecule, creating and expelling a water molecule in the process—are promising candidates, because water molecules readily split the anhydride linkages in the reverse of the process that created them, yet the polymer molecules can still be water-repellent in bulk. By varying the ratios of the components, surface-eroding polymers lasting from one week to several years have been synthesized.

These polymer disks are now being used experimentally as a postoperative treatment for brain cancer. The surgeon implants several polyanhydride disks, each about the size of a quarter, in the same operation in which the brain tumor is removed. The disk contains powerful cell-killing drugs called nitrosoureas. Nitrosoureas are normally given intravenously, but they are effective in the bloodstream for less than an hour. Unfortunately, nitrosoureas are indiscriminately toxic, and this approach generally damages other organs in the body while killing the cancer cells. But placing the drug in the polymer protects the drug from the body, and the body from the drug. The nitrosourea lasts for approximately the duration of the polymer—in this case, nearly one month. And the eroding disk delivers the drug only to its immediate surroundings, where the cancer cells lurk.

The polymer degradation method of drug delivery is making good progress toward approval by the Food and Drug Administration.

tain applications. (See the vignette "Seasickness Patches.") Recently, more sophisticated technologies have emerged, such as electrotransport systems, whereby the drug is driven from a reservoir under the influence of an electric field. Such systems are being developed predominantly for transdermal drug delivery.

There are many challenges in designing polymers for controlled-release applications. These polymers must be biocompatible, pure, chemically inert, nontoxic, noncarcinogenic, highly processible, mechanically stable, and sterilizable. The polymers in use today in drug delivery are also mostly borrowed from the chemical industry and in many cases lack the exact required properties. Novel polymers designed and synthesized to provide optimal properties and characteristics will be required to take full advantage of the emerging technologies described above.

A number of controlled-release products are on the market in the United

SEASICKNESS PATCHES

The venerable drug scopolamine, found in henbane and deadly nightshade, is perhaps the most effective short-term preventer of motion sickness. Unfortunately, scopolamine does not stay in the blood long. Because the drug must be taken at short intervals, the possibility of accidental overdose—with its side effects of drowsiness, blurred vision, hallucinations, and disorientation—is increased. This tended to limit the drug's popularity as a seasickness preventative. A way to deliver a constant low dose to the bloodstream for hours on end needed to be found.

A polymer-based "transdermal patch" proved to be the answer. The thickness of a playing card and less than three-eighths of an inch in diameter, the patch is applied like an adhesive bandage and does not break the skin. The skin behind the ear is the most permeable, and from there the scopolamine rapidly diffuses into the blood vessels just below the surface.

The patch consists of several laminated layers of different polymers, each one designed for a different function. The topmost layer is a polyester film, colored to match the skin. Adhering to the polyester's underside is a film of vapor-deposited aluminum to protect the drug from sunlight, evaporation, and contamination. Then comes a polymer adhesive that binds the aluminum to the rest of the patch. The next layer, the reservoir, is made of a polyisobutylene skeleton filled with mineral oil that contains a 72-hour supply of the drug in a special skin-permeable formulation. Between the reservoir and the skin is a polypropylene membrane riddled with microscopic pores. The pores are just the right size to ensure that the drug seeps out at a rate less than it can be absorbed by the most permeable skin. This feature ensures a constant dose rate, regardless of the skin's permeability. The patch's bottom layer is an adhesive formulation of polyisobutylene and mineral oil. This mineral oil also contains the drug, so that it saturates the skin as soon as the patch is applied and minimizes the time lag before the scopolamine takes effect. (Even so, it generally takes about 4 hours to kick in.) The adhesive layer is protected before use by a peel-off backing of siliconized polyester.

The transdermal patch technology transformed an otherwise unmanageable drug into the most effective motion sickness treatment available, and one good for three days. Yet this seemingly simple patch—a glorified sticker/Band-Aid—employs at least six layers of carefully chosen polymers, each of which has a specific function and each of which must be compatible with the neighboring materials. Designing and testing the patch required attention to complex issues of drug dosage and behavior as well as the challenge of fabricating a pharmaceutical product in a radically different and untried form.

States, and a larger number are in development. Table 2.1 lists a few of the available products as examples of drug delivery systems, along with the mechanisms of release and polymers used as major components of the rate-controlling element.

TABLE 2.1 Examples of Currently Available Controlled-release Drug Delivery Systems

Product	Drug	Delivery Route	Major Rate-controlling Polymer	Mechanism of Release	Indications for Use
Procardia XL®	Nifedipine	Oral	Cellulose acetate	Osmotic	Hypertension and angina
Duragesic®	Fentanyl	Transdermal	Ethylene vinyl acetate	Diffusional	Chronic pain
Proventil/Repetabs®	Albuterol	Oral	Acacia/carnauba wax	Erosional	Asthma
Estraderm®	Estradiol	Transdermal	Ethylene vinyl acetate	Diffusional	Hormone replacement
Norplant®	Levonorgestrel	Implant	Silicone rubber	Diffusional	Contraception
Catapres-TTS®	Clonidine	Transdermal	Polypropylene	Diffusional	Hypertension
Zoladex®	Goserelin	Implant	Polylactide-glycolide	Erosional	Prostate cancer

SOURCE: Compilation of information from *Physicians' Desk Reference* (1994).

Biological Polymers

Biopolymers Versus Synthetic Polymers

The volume of biopolymers in the world far exceeds that of synthetic macromolecules. Biological polymers include DNA, RNA, proteins, carbohydrates, and lipids. DNA and RNA are informational polymers (encoding biological information), while globular proteins, some RNAs, and carbohydrates serve chemical functions and structural purposes. In contrast, most synthetic polymers, and fibrous proteins such as collagen (which makes up tendon and bone) and keratin (which makes up hair, nails, and feathers), are structural rather than informational or chemically functional. Structural materials are useful because of their mechanical strength, rigidity, or molecular size, properties that depend on molecular weight, distribution, and monomer type. In contrast, informational molecules derive their main properties not simply from their size, but from their ability to encode information and function. They are chains of specific sequences of different monomers. For DNA the monomers are the deoxyribonucleic acid bases; for RNA, the ribonucleic acid bases; for proteins, the amino acids; and for carbohydrates or polysaccharides, the sugars. The paradigm in biopolymers is that the sequence of monomers along the chain encodes the information that controls the structure or conformation of the molecule, and the structure encodes the function. An informational polymer is like a necklace, and the monomers are like the beads.

For RNA and DNA, there are 4 different monomers (beads of different colors). Information is encoded in the sequence of bead colors, which in turn controls the sequence of amino acids in proteins. There are 20 different types of amino acid monomers; in the necklace analogy, there are 20 different colors of beads. A globular protein folds into one specific compact structure, depending on the amino acid sequence. This balled-up shape, or structure, is what determines how the protein functions. The folding of the linear structure produces a three-dimensional shape that controls the function of the protein through shape selection.

Except in special cases, synthetic polymer science does not yet have the precision to create specific monomer sequences: polymers can be synthesized as homopolymers, chains composed of only a single type of monomer, or simple block copolymers, where the monomers repeat only in the simplest patterns, AAA BBB AAA BBB, or random sequences. But the ability to synthesize specific monomer sequences by a linear process would have extraordinary potential. For example, it is the ability to create specific monomer sequences that distinguishes biological life forms, and the corresponding complex hierarchies of structure and function, from simpler polymeric materials. Hence one of the most exciting vistas in polymer science is the prospect of creating informational polymers through control of specific monomer sequences. The present state of the art is exemplified by Merrifield-type syntheses, in which polypeptides are synthesized one amino acid at a time on an insoluble support composed of polypeptides and polynucleic acids. This method is limited to preparation of short chains (less than 50 amino acid groups) and small quantities. Techniques that allow similar controlled synthesis on a much larger scale would be revolutionary.

The study of informational polymers aims to determine the specific shapes of biological polymers at atomic and nanometer resolution, the relationship between structure and function, and how the structure and function arise from the underlying interatomic forces of nature. Because these are the same goals as in the study of synthetic polymers, the topics of biomaterial-related polymer science and engineering cut across all the areas of this report.

Biopolymers in Molecular Recognition

A major goal of science is to learn how one molecule binds, recognizes, and interacts with another molecule. If the principles that control the binding and recognition events were understood, we could design activators for biomolecules and drugs, understand biological regulation, and improve separation methods. Major strides are occurring in the following areas: (1) Structures of biomolecule complexes are becoming available, including antibody-antigen complexes, ligands with proteins or DNA or RNA, proteins with DNA, and viruses and ribosome assemblies. (2) Computer programs are being developed to allow databases to be searched to find promising binding candidate molecules. (3) Combinatoric peptide templates, which are arrays of very large numbers of different peptides attached to surfaces, are allowing rapid screening of large numbers of possible binding agents for a specific bioprobe and have the potential to speed up drug design by many orders of magnitude.

Biopolymers in Biological Motion

The cellular machinery for motion is complex and varied. For example, some bacteria are propelled by their flagellae, which act like small rotors. Vertebrate muscle motion depends on the actomyosin system, whose major components are the proteins actin and myosin. The myosin fibers move along the actin fibers, powered by cellular processes involving adenosinetriphosphate (ATP). The exact motions of the myosin molecules are not yet understood. The structures of both the actin and the myosin proteins have recently been determined by crystallography. New methods have recently been developed that probe forces and motions, including a mobility assay for watching the motions of muscle and related proteins under the microscope, "optical tweezers" for measuring forces, and electron spin resonance experiments for detecting conformational changes. Major advances are happening very rapidly now.

Bioelastomers

To obtain high elasticity and the desirable properties it imparts, polymers are needed that have high chain flexibility and mobility. This need has led both nature and industry to choose polymers with small side chains, little polarity, and a reluctance to crystallize in the undeformed state. Rubberlike elasticity arises from the flexible chains interconnecting the cross-linking of polymer chains. The cross-linking carried out in nature is more sophisticated than the cross-linking used in the production of elastomers in the laboratory. In biological systems, cross-links are introduced at specific amino acid repeat units and are thus restricted both in their number and in their locations along the chain. Furthermore, they may be carefully positioned spatially as well, by being preceded and succeeded along the chain by rigid alpha-helical sequences. If we had nature's ability to control network structure, it would be possible for us to design materials with better mechanical properties. For example, many bioelastomers have relatively high efficiencies for storing elastic energy through the precise control of cross-link structure. A desirable advanced material would be an elastomer with low energy loss. Such a material would have the advantages of energy efficiency and fewer problems from degradation resulting from the heat buildup associated with incomplete recovery of elastic energy. Another desirable advanced material would have high toughness, which may be obtained by exploiting non-Gaussian effects that increase the modulus of an elastomer near its rupture point. Some work on bioelastomers suggests that toughness may be controlled by the average network chain length and the distribution about this average. There have been attempts to mimic this synthetically by end-linking chains of carefully controlled length distributions, but much more should be done along these lines.

Biocomposites

Biocomposites are usually composed of an inorganic phase that is reinforced by a polymeric network. The various types of biocomposites found in nature, such as bone, teeth, ivory, and sea shells, differ from synthetic analogs in one or more important respects. First, the hard reinforcing phase in biocomposites is frequently present to a very great extent, in some cases exceeding 96 percent by weight. Second, the relative amounts of crystallinity, morphology, and crystallite size and distribution are carefully controlled. Moreover, the orientation of crystalline regions is generally fixed, frequently by the use of polymeric templates or epitaxial growth. Third, instead of a continuous homogeneous phase, a gradation of properties in the material is obtained by either continuous changes in chemical composition or physical structure. Finally, larger-scale ordering is often present, for example, in complex laminated structures, with various roles being delegated to the different layers present.

The differences cited above are achieved in biocomposites by nature's use of processing techniques that can be entirely different from those that have been used for synthetic composites. Until recently, in the methods used for synthetic composites, the two or more phases have generally been prepared separately and then combined into the composite structure. Occasionally, some chemistry is involved, but it is, typically, relatively unsophisticated, for example, the curing of resin in a fiberglass composite.

More intelligent approaches are now being used to design materials, particularly those required to have multifunctional uses. In particular, the types of chemical methods that predominate in the construction of biocomposites are being used increasingly by materials scientists. These syntheses are carried out in situ, with either the two phases being generated simultaneously or the second phase being generated within the first. The generation of particles or fibers within a polymer matrix can avoid the difficulties associated with blending agglomerated species into a high-molecular-weight, high-viscosity polymer. The dispersed phase can be present to much greater extents, and much work could be done on the problem of using the polymeric matrix to control its growth. It may also be possible to avoid geometric problems, such as the alignment of fibrous molecules packed to high densities either because of their response to flow patterns or because of their inherent symmetry. Such anisotropy can be disadvantageous in that it leads to strengthening the material in some directions, but at the cost of weakening it in others. When such molecules are grown within an already formed matrix, however, essentially random isotropic packing can be obtained. The shell of the macademia nut is an excellent example of this type of reinforcement. In it, bundles of cellulose fibers are present in structures having considerable alignment. The composite is, thus, random and isotropic at larger scale, and this is the source of its celebrated toughness. Similar arrangements occur in some liquid crystalline polymers, but there is little correlation between the axes of different domains, and nothing has been done yet to mimic this type of composite material. In the case of chemically based methods, the competition between the kinetics of the chemical reactions and the rates of diffusion of reactants and products can also be used to advantage, for example, in the formation of permanent gradients. This approach is yet another opportunity to exploit nature's ideas.

Overlap Between Structural Biology and Polymer Science

The above exciting areas involve considerable overlap between biomaterials and polymer science. Polymers and biopolymers have a number of common elements, including the problems of understanding molecular conformations as the basis of underlying chemical events, the subtle driving forces, often largely entropic, and considerable overlap in the experimental and theoretical methodologies. Despite the considerable overlap in problems and methodologies in polymer science and those in many areas of biology, there is still little crossover in research and background knowledge between these fields. Both fields would benefit substantially from more crossover and cross-education.

INFORMATION AND COMMUNICATIONS

The past half-century has witnessed an explosion in electronics and communications. Our world has been transformed as the transistor-based technologies have given rise to new modes of information storage, processing, and transmission, vital to enhanced productivity, improved health care, and better transportation systems. These technologies are abundantly evident as supermarket scanners, fax machines, word processors, automatic teller machines, and many other "essentials" of modern life. Silicon and software are legitimately most clearly associated with these advances, but other materials, including polymers, play an essential supporting role, which is growing in importance. Owing to their high performance, manufacturing flexibility, quality, and low cost, polymers are key factors. The role of polymers is predicted not only to increase in quantitative terms, but also, more importantly, to extend into new areas in which polymers have not been employed in the past.

Historically, polymeric materials have been applied mainly as insulators and packaging. These uses often involve substantial quantities of material, for example, several hundred million kilograms for cable production annually, and they will remain important for the long-term future. In these applications, polymers offer ease and economy of manufacture, tough, durable mechanical properties, and excellent dielectric properties (i.e., low dielectric constant and loss). Polymers are unlikely to be challenged in these areas. Polyethylene is consistently the material of choice for most communication and power cables, but fluorinated and other polymers are becoming increasingly important for special applications, such as inside wiring where flammability considerations are paramount.

Over the last 20 years, polymers (and other organic materials) have been developed that exhibit electrical and optical properties that were formerly found only in inorganic materials. Polymers have been found that are piezoelectric, conduct electricity electronically, exhibit second- and third-order nonlinear optical behavior, and perform as light-emitting diodes. Optical wave guides, splitters, combiners, polarizers, switches, and other functional devices have been demonstrated. In addition, lithographic pattern formation by the interaction of polymers with ultraviolet (UV) light and other forms of radiation has been carried to amazing levels of resolution and practicality and is the basis for fabrication of integrated and printed circuits of all kinds.

In this section, some of these more exotic properties of polymers are briefly described. For many of these materials, applications are only now being developed. It is likely that the new applications will have specialty niche markets, unlike the massive present market of commodity polymers. The economic factor driving their production will be quality; small quantities of carefully controlled materials will be produced at high unit costs. These products will be sold by function, not weight.

Polymer Dielectrics for Electronics

Organic polymers play a crucial role as insulating materials in electronics. The most visible applications are in silicon chip encapsulation and in dielectric layers for printed circuit boards (PCBs). (Further details on PCBs are given in the vignette "Printed Circuit Board Materials.") Encapsulation of chips is accomplished through transfer molding in which the chip, attached to its metal lead frame, is covered entirely with plastic, leaving only the ends of the lead frame connectors exposed for connection to printed wiring board (PWB) pads. The polymer employed is usually an epoxy (novolac) that is highly loaded with silica powder to reduce the coefficient of thermal expansion. Differences in thermal expansion between chip and encapsulant create large stresses on cooling from mold temperatures and as the temperature of the assembly is cycled in testing and in use. Encapsulation is mainly for mechanical and chemical protection of the chip and the lead frame and thus facilitates handling for automatic assembly. Materials and processes have been developed to a high degree of sophistication. High mechanical strength is achieved with the smallest external dimensions.

Printed circuit boards are layered structures of patterned copper connection paths ("wires") placed on a polymer substrate. The width of the "wires" is typically 100 to 200 micrometers (μm). Polymers employed include epoxies, polyesters, fluoropolymers, and other materials, but glass-reinforced epoxies (usually bisphenol-A based) are by far the most widely used. Metal patterns are defined photolithographically and plated to the desired thickness, and the layers are then piled up and cured in a press. Circuits with more than 40 copper layers (signal, power, and ground) have been produced commercially. Connection to the inner layers is made through "via" holes that are copper plated. One supercomputer was marketed in which all of the electronics was placed on a single multilayer circuit board. The materials and process control requirements are challenging, and the functional end-product is worth a great deal.

In some cases a finer form of interconnection is needed, and this is provided by hybrid circuits based on an alumina substrate (with "wire" widths of about 75 μm) and multichip modules (MCMs) usually built on a silicon wafer (with "wire" widths in the 10- to 50-μm range). MCMs represent the leading edge of interconnection technology, and they are used when the time of transit of signals from chip to chip is an important limitation on the processing speed of the electronic system. The speed of light is the ultimate barrier, and consequently it is essential to employ dielectrics that have the lowest practical dielectric permittivity. This is an area in which polymers offer substantial advantages over inorganic dielectrics.

PRINTED CIRCUIT BOARD MATERIALS

Practically any twentieth-century gadget you can think of, from the cheapest clock-radio to the most expensive mainframe computer, has its electronic guts mounted on printed circuit boards. These "boards"—actually fiberglass cloth impregnated with a brominated epoxy polymer resin—got their name because the electronic components on them are wired together by thin copper ribbons deposited directly onto the boards, like ink on paper. The idea that bulky, plastic-clad copper wires could be replaced by ribbons of bare metal on an insulating background was one of the fundamental breakthroughs of the electronics revolution of the 1960s. Since then, printed circuit board manufacture has grown into a $20B-per-year business.

Printed circuit board substrates are an example of a "composite material"—a multicomponent material that performs better than the sum of the properties of its individual components. The chemical structures of such a material's components, and their relative proportions, can be tailored to provide just the right set of properties for a given application. In this case, the material has to be not only lightweight and strong but also an electrical insulator, which rules out the use of metal sheets. The material must also be fracture-resistant, so that it can be cut to shape or drilled without cracking. And the material must be thermally stable—some of the newest, high-technology computer chips give off a lot of heat. Where such a chip is mounted, the board can be exposed to temperatures of up to 121°C. The board has to handle such a hot spot without melting. The board also has to be flame retardant, so that an electrical short does not become a conflagration that wipes out a lot of expensive hardware. In this composite material, the glass-fiber cloth gives the board its lightweight strength, while the brominated epoxy resin eventually becomes a rigid, three-dimensional network that gives the board the necessary stiffness, fracture resistance, and other properties.

The manufacturing process starts with a roll of glass-fiber cloth. Carefully adjusted tension rollers feed the cloth at a precisely determined rate through a bath of the resin, which has been dissolved in a solvent. The resin-impregnated cloth then wends its way over other rollers and through a series of ovens to evaporate the solvent. The heat and a catalyst also "cure" the resin—promoting the chemical reactions that harden it into a tough, durable solid. Several layers of partially cured cloth can be laminated together before further curing to make an even stronger circuit board. Finally, the cured board, now as stiff as its namesake, is sawn up into the individual circuit boards.

Circuit board substrate materials have evolved over the years. New epoxies are now being used to improve dimensional control. Alternative polymer matrices are used for applications demanding high-temperature performance. Polymers are also being used for the reinforcing fibers themselves. Printed circuit boards, the key interconnection medium for electronics, depend critically on polymers and their composites.

By far the most research and development on materials for MCM dielectric layers has gone into polyimides, and most existing applications are based on polymers of this family. Great strides have been made in achieving the demanding property mix required through careful tailoring of the monomer chemistry. Improved adhesion, lower dielectric constant, reduced sensitivity to moisture, higher thermal stability, and other properties have been improved greatly. The in-plane coefficient of thermal expansion was reduced and adjusted to the range of silicon, metals, and ceramics. Most major electronics companies manufacture MCMs based on polyimides.

In spite of the extent of commitment to polyimides, it has proved difficult to achieve all the desired properties in a given composition. Other polymer dielectrics are in use, and new materials are under consideration. For example, commercial MCMs are manufactured by one electronics systems provider based on a proprietary epoxy-acrylate-triazine polymer that is photodefinable. Sample MCMs have been produced based on a benzocyclobutene (BCB) polymer dielectric. In spite of the large experience base with the polyimide materials, the newer polymers have advantages and offer attractive alternatives. All of the candidates are glassy polymers. The dielectric constants may be compared as follows:

alumina	9
glass ceramics	4-5
fused silica	4
polyimides	3-4
triazine	2.8
BCB	2.7

In the final analysis, the choice of materials will be based on the sum of property advantages and processing practicality. Polymers offer the lowest dielectric constants and the thinnest "wires."

Lithographic processes and associated technologies have advanced to the point that semiconductor device cells and conductor lines (i.e., the on-chip "wires") are so small (less than 1 μm) and the switching times are so fast that the continual increase in performance traditionally derived from a combination of improvements in device structure and reduction of device dimensions cannot be fully realized. This is owing to the fact that the propagation of signals through the wiring on the chip (and in the module) is becoming the dominant limitation on processor cycle time.

The velocity of pulse propagation in these structures is inversely proportional to the square root of the dielectric constant of the medium. Hence, reductions in the dielectric constant translate directly into improvements in processor cycle time, in part because of the speed of propagation. In addition, the distance between signal lines is dictated by noise issues or "cross-talk" that results from induced current in conductors adjacent to active signal lines. A reduction of the insulator dielectric constant permits moving the signal lines closer together, allowing designers to reduce the length of conductor lines and thereby improve cycle time.

Performance demands on polymers incorporated as permanent parts of the chip structure are even more stringent than the requirements for MCMs and PCBs. Insulating materials in chip applications must be able to withstand the very high temperatures associated with the processes used to deposit metal lines and to join chips to modules. At a minimum, they must withstand soldering temperatures without any degradation or outgassing. They must have thermal expansion coefficients that are closely matched to that of silicon. Silica meets all of the requirements extremely well, and this would continue to be the material of choice were its dielectric constant not so high.

While much attention has been given to polymers with very low permittivities, there is an increasing need for high-permittivity polymers in capacitor applications. The rational design of polymers having high (ε > ca. 10 to 15) permittivities and low loss has not been pursued, and this represents an attractive opportunity for joint efforts in molecular modeling and polymer synthesis.

Clearly, organic polymers currently play a critical role as insulators in electronic devices and systems. Continued success in the development of new generations of these critical dielectric materials depends on close interactions between the microelectronics and the chemical communities, a relationship that is not in evidence in the United States. New partnerships are needed if we are to maintain competitiveness in this vital industry.

Conducting Polymers and Synthetic Metals

Organic materials are generally insulators or, in other words, poor conductors of electricity compared with metals and semiconductors. Electrical conductivity in metals and semiconductors arises from the delocalized electrons of the system, and they are best described by "band theory." In these terms, the organic materials have localized electrons because there is a large energy gap between the most energetic electrons and the conduction band. It has long been known that conjugated systems, that is, linear systems with alternate double and single bonds, should have delocalized electronic states, but it was only in 1977 that polyacetylene was shown to exhibit true metallic conductivity. Earlier, in the 1960s, low-molecular-weight organics had been shown to behave as semiconductors (e.g., TCNQ) and metals (e.g., TCNQ:TTF). Those discoveries stimulated a large amount of research leading to the preparation of many new molecular metals and understanding of the nature of this new class of materials.

An organic polymer that possesses the electrical and optical properties of a metal while retaining the mechanical and processing properties of a conventional polymer, is termed an "intrinsically conducting polymer" (ICP), more commonly known as a "synthetic metal." The properties of these materials are intrinsic to a "doped" form of the polymer.

The concept of "doping" is the unique, central, underlying, and unifying theme that distinguishes conducting polymers from all other types of polymers. In the doped form the polymer has a conjugated backbone in which the π-system is delocalized. During the doping process, a weakly conducting organic polymer is converted to a polymer that is in the "metallic" conducting regime (up to 10^4 siemens per centimeter [S/cm]). The addition of small (usually <10 percent) and nonstoichiometric quantities of chemical species results in dramatic changes in the properties of the polymer. Increases in conductivity of up to 10 orders of magnitude can be readily obtained by doping. Doped polyacetylene approaches the conductivity of copper on a weight basis at room temperature. Doping is reversible. The original polymer can be recovered with little or no damage to the backbone chain. The doping and undoping processes, involving dopant counter ions that stabilize the doped state, may be carried out chemically or electrochemically. By controllably adjusting the doping level, a conductivity anywhere between that of the undoped (insulating or semiconducting) and that of the fully doped (metallic) form of the polymer may be obtained. Conducting blends with nonconducting polymers can be made. This permits the optimization of the best properties of each type of polymer.

All conducting polymers (and most of their derivatives), including polyacetylene, polyparaphenylene, poly(phenylene vinylene), polypyrrole, polythiophene, polyfuran, polyaniline, and the polyheteroaromatic vinylenes, undergo either p- and/or n-redox doping by chemical and/or electrochemical processes during which the number of electrons associated with the polymer backbone changes. P-doping involves partial oxidation of the π-system, whereas n-doping involves partial reduction of the π-system. Polyaniline, the best-known and most fully investigated example, also undergoes doping by a large number of protonic acids, during which the number of electrons associated with the polymer backbone remains unchanged.

Appropriate forms and derivatives of many conducting polymers, especially those involving polyaniline and polythiophene, are readily solution processible into freestanding films or can be spun into fibers that even at this relatively early stage of development have tensile strengths approaching those of the aliphatic polyamides. Blends of a few weight percent of conducting polymers with aromatic polyamides or polyethylene can exhibit conductivities equal to, or even exceeding, the conductivity of the pure conducting polymer while retaining mechanical properties similar to those of the host polymer. In addition, pure conducting polymers and their blends can be oriented by stretching to produce highly anisotropic electrical and optical properties.

The thermal, hydrolytic, and oxidative stability of doped forms of pure conducting polymers varies enormously from the n-doped form of polyacetylene, which undergoes instant decomposition in air, to polyaniline, which has sufficient stability in air at 240°C to permit blending and processing with conventional polymers. The oxidative and hydrolytic stability is significantly increased when the conducting polymer is used in the form of blends with conventional polymers. Clearly, research to improve the stability of conducting polymers is essential to commercial applications in the future.

Polyaniline is currently the leading conducting polymer used in technological applications and is commercially available in quantity. Polypyrrole and derivatives of polythiophene and poly(phenylene vinylene) also have significant potential technological applications. Rechargeable polyaniline batteries and high-capacity polypyrrole capacitors are in commercial production.

Ionically conducting polymers are now being used in batteries and electrochromic displays. However, even though conductivities of greater than 10^{-3} S/cm are now achievable with gel electrolytes, the goal of preparing single-ion (and specifically cation) conductors with comparable conductivities has remained elusive. Tight ion pairing between Li^+ and polymer-bound anions (usually sulfonates) is responsible for the significantly lower conductivities. Also, new approaches for the synthesis of polymer electrolytes as thin films directly on electrodes (via, for example, photopolymerization) are needed to complement novel multilayer battery fabrication technology. Along these lines, a key goal is the design of multifunctional polymers capable of transporting only cations, stabilizing a battery system against overcharging, and exhibiting low reactivity at alkali metal and metal oxide electrodes. Perhaps most important, electrode-polymer electrolyte reactions need to be examined from a fundamental point of view because these represent a major problem for battery cyclability and overall stability.

Polymer Sensors

The field of sensors is diverse, reflecting our need to control increasingly complex systems—including environments, processes, equipment, vehicles, and biomedical procedures—that are characterized by high levels of automation. The key to the success of such automated systems is the measurement technology, which demands rapid, reliable, quantitative measurement of the required control parameters. These parameters include temperature, pressure, humidity, radiation, electric charge or potential, light, shock and acoustic waves, and the concentrations of specific chemicals in any environment, to name just a few. Obviously, the types of sensors that are applied to such wide-ranging measurements are quite varied in type and principle of operation. Nevertheless, polymers play a significant role as enabling active materials for the design of sensors that are extending current limitations of sensitivity, selectivity, and response time.

A great deal of sensor research and development is focused on tailoring polymeric materials for applications in the chemical and biomedical fields. First, polymers can be functionalized through the incorporation, in their syntheses or thereafter, of moieties that respond in some detectable way to the presence of the chemical to be analyzed. For example, polymers that are modified to bind dyes that respond to blood chemistry (oxygen, carbon dioxide, acidity) or to immobilize enzymes that produce reactions with substances of biological interest, such as glucose, are used to construct biosensors for in vivo application.

Another polymer property used in such sensors is permeability. The polymer allows diffusive transport of the chemical to the immobilized functionality to enable interaction and subsequent detection of the reaction products. When increased transport kinetics are required, the polymer may be fabricated in a porous state or may be engineered to swell or expand in the medium in which the sensor is immersed, such as water. In other cases, polymers may be engineered as a controlled-release material, supplying reagents to the surrounding medium for local detection.

The polymer properties described here are being used in the development of fiber-optic chemical sensors. These sensors employ dye molecules incorporated into transparent polymers that form either part of the fiber structure or part of an active element, termed the "optrode," located at the terminus of the fiber. The sensors may incorporate either absorbing or fluorescent dyes for detection of specific chemical species. Light injected into the fiber, at a location remote from the chemical environment being probed, interacts with the dye and is absorbed or produces fluorescence. When a chemical species permeates the polymer and alters the absorption or fluorescence of the dye, the light output of the fiber returning from the optrode is altered in a quantitatively detectable manner.

Chemically modified electrode sensors rely on the measurement of electrical potentials produced by selective electrochemical reactions involving the chemical species to be determined. The development of thin polymer coatings to chemically modify the electrodes is an important topic of research in this field. The polymers are chemically and physically modified to concentrate electroactive sites at the electrode surfaces, to provide large ion and electron mobility, and to ensure a stable environment for the desired electrochemical reactions. Especially promising areas of investigation include the development of such sensors for determination of specific ions and products of biochemical reactions with enzymes or antibodies immobilized in the polymer film.

A related sensor type in the chemical and biomedical fields is the microsensor based on integrated solid-state electronic devices, for example, CHEMFETS. These sensors incorporate chemically sensitive polymer films placed in contact with the gate of a field effect transistor on a transducing silicon chip. The electrical current output of the device is modulated by the chemical environment at its surface. The polymer films are tailored in their chemical and physical properties to optimize specific solubility interactions and/or chemical activity with the substance to be sensed, thereby controlling the sensitivity and selectivity of the sensor. Polymers used for this purpose must often be deposited and patterned using the standard photolithographic techniques of the semiconductor industry and then undergo further chemical modification in order to impart necessary properties to the device. Integration of signal processing functions on the sensor chip and on-chip sensor arrays for simultaneous determination of a range of chemical entities are key aspects of the development of this sensor type. These sensors are being applied to analyses ranging from ionic species to gaseous and liquid chemicals and biochemical substances. Sensors that can be implanted in the body are a major goal. Much effort is being devoted to glucose sensors that would allow insulin pumps to respond to a diabetic person's time-dependent need for this vital hormone.

An important extension of the solid-state microsensor makes use of electronically conducting conjugated polymers. The electronic conductivity of these materials is modulated over several orders of magnitude by interaction with a variety of chemicals. The polymers are deposited on electrodes or solid-state devices by electrochemical polymerization, and dopants are simultaneously incorporated in the polymerization process to enhance conductivity and chemical activity. Sensors of this type have been applied primarily to the detection of gases (such as ammonia, nitrogen dioxide, and hydrogen sulfide) and ions.

Specific polymers, called electrets, have the ability to store electrical charges or to be electrically poled so that they retain a permanent polarization. These polymers can be fabricated into specific structures in which their deformation or movement produces electrical signals that can be resolved. Electret materials, best exemplified by fluorinated polymers such as poly(tetrafluoroethylene), can be fabricated into films, charged, and used to construct condenser-type acoustic transducers (electret microphones). Ferroelectric polymers, such as poly(vinylidene fluoride), can be poled by applying a strong electric field, and then used to construct acoustic, pressure, or thermal sensors. They are applied in pyroelectric detectors, hydrophones, ultrasonic transducers, shock wave sensors, and tactile sensors for robotics. Often composites of these polymers with piezoelectric ceramics are used to provide enhanced performance.

Polymers that emit light when exposed to ionizing radiation or high-energy particles are used as the active elements in radiation detectors (scintillation detectors). These polymer systems have advantages over liquid scintillation detectors because of their ease of fabrication and ruggedness with comparable sensitivity.

Resist Materials

For the last decade, the microelectronics industry has been engaged in a race to shrink the dimensions of semiconductor devices. The result of this effort is the continued improvement in the price-to-performance ratio of microelectronic devices and the myriad products that are produced from them (see the vignette "Resists and Micromachines"). The market for silicon hardware will exceed

RESISTS AND MICROMACHINES

Imagine a tiny robot—a micromachine—the size of a red blood cell, swimming through the arteries of a stroke victim until it reaches the blood clot in the victim's brain. The micromachine drills through the clot, restoring blood flow. The parts for such robots might one day be built using the same polymers that are used to stencil the incredibly complex pattern of an integrated circuit onto a silicon chip. These polymers, called resists, react chemically when exposed to ultraviolet light, X-rays, or other energetic electromagnetic radiation. One polymer commonly used as a resist, poly(methyl methacrylate), is better known to most people as Plexiglas.

Tiny gears, for example, have already been made. The process starts with a blank wafer of silicon, to which a thin layer of titanium has been applied as a sort of frosting. A layer of resist, as thick as the gear is supposed to be (usually about a few microns thick, or much less than the thickness of a human hair), is applied to the titanium surface. The wafer is then bombarded with X-rays that have passed through a gold mask with many gear-shaped holes cut in it. Wherever the X-rays hit the wafer, the resist molecules become soluble. Wherever the wafer is shielded by the mask, the resist does not react and remains insoluble. Washing the wafer in the solvent mixture leaves a gear-shaped hole in the resist to use as a form. The form is filled by electroplating copper into it—the titanium layer on the wafer is connected to a negatively charged electrode, and the wafer is immersed in a solution of copper ions. The copper deposits itself on the exposed titanium, but not on the resist, which does not conduct electricity. Once the resist is removed by an aggressive solvent or oxygen plasma, the free-standing copper gear remains on the titanium. Dunking the wafer, gears and all, into a bath of hydrofluoric acid dissolves the surface titanium, freeing the gears from the wafer.

Employing the pattern-making capability of polymer resists, it is feasible to make metal gears of the order of 1 micrometer in diameter. Living cells are tens of micrometers across, and a small blood vessel is about 50 micrometers in diameter. Thus, fairly elaborate machines appear possible, although other parts and assembly will require a great more development effort.

$100B and the market for systems based on silicon will exceed $1,000B in the next few years, opening a highly competitive international market (Sze, 1988).

The processes employed to manufacture the silicon hardware are intrinsically dependent on the polymeric materials that are used to define the patterns required for the many layers of circuitry. In a typical process, several hundred steps are required to produce a wafer containing hundreds of chips. About two-thirds of these steps are devoted to pattern formation, a form of lithography. In the process as practiced by the semiconductor industry, the silicon wafer on which the devices (e.g., the individual transistors and other elements) are to be fabricated is coated with a thin film of a material called a resist. Pattern-wise exposure of the resist to radiation of the appropriate wavelength results in a radiation-induced chemical reaction in the resist film, which renders the exposed areas more soluble in some developer solvent (positive tone imaging) or less soluble (negative tone imaging). The pattern is formed by passing the radiation through a mask, which is similar to a stencil, which blocks the radiation in areas in which no reaction is desired. The result is a relief image consisting of regions of resist and regions of bare circuit. These relief images in the resist allow the underlying substrate to be processed selectively in those areas where the resist has been removed. The processes involved include etching, metal deposition, ion implantation, and oxidation of silicon.

Virtually all production of semiconductor devices is accomplished by exposing the resist film to UV radiation through a projection lens system analogous to the familiar slide projector, although exposure tools shrink the projected image rather than expand the image of the slide. Present systems employ UV radiation with a wavelength of 365 nanometers (nm), but 248-nm systems are being introduced. Electron beam and X-ray radiation offer alternatives for the future. Each change in wavelength and radiation type requires development of new polymeric resist materials.

Owing to the high cost of the exposure tools, it is important that the throughput of the machines (e.g., the rate at which they can produce exposed wafers) be as high as possible. The amount of light available at 248 nm is only one-tenth that provided by the older machines operating in the near UV. Therefore, the feasibility of moving to deep UV was entirely dependent on the ability of chemists to develop new generations of polymeric resists that are as much as 100 times as sensitive as resists formerly used. These new resists derive their high sensitivity from exploitation of an acid-catalyzed reaction that converts an insoluble moiety to one that is soluble. Exposure converts a neutral substance into an acid, thereby generating a latent image of the mask. The resulting films are then baked to provide the activation energy necessary to start the catalytic reaction in which the acid generated upon exposure facilitates a reaction of the resist polymer (i.e., at a side group) to convert it to a form that is soluble in the developer. The radiation-created catalyst can convert many polymer groups, giving rise to the "chemical amplification" made necessary by the scarcity of deep-UV photons. Although there are many other factors involved in moving from near- to deep-UV lithography, both chemical and other, development of the chemically amplified polymeric resists was an essential contribution.

As the lateral dimensions of devices shrink, the width of the resist images required to define their component structures must shrink also. The thickness of the resist film does not shrink, however, owing to the necessity of being pinhole free and robust in subsequent processing (e.g., ion implantation or etching). Thus, the aspect ratio of the relief structure is increasing and could be as high as five by the end of the decade. This is a very demanding requirement that will require a significant advance in resist technology.

One promising approach to the production of high-aspect-ratio imaging at small dimensions is "top surface imaging." In this process the resist film is formulated to be opaque to the exposure radiation, and chemical transformation occurs only on the top surface of the resist. The transformation is designed to produce selective reactivity with an organometallic reagent such that only the exposed surface incorporates the organometallic. If silicon is incorporated, subsequent anisotropic oxygen etching of the film results in rapid formation of a thin layer of silicon dioxide in the areas that reacted with the reagent. This thin oxide layer protects the polymer beneath while the unexposed, unprotected polymer areas are etched away by the oxygen plasma. The products of the etching are gaseous and are pumped away. The aspect ratio of the image produced by this process is dependent on the anisotropy of the oxygen etching process. Aspect ratios exceeding five in polymer relief images have been achieved by this method. Although many features of the top surface imaging procedure remain to be worked out, it is a promising method.

Some solution must be found that will provide support for the research and development required to produce these materials that are so critical to the continued advance of semiconductor technology. Sematech, the U.S. (industry-government supported) consortium is investing at some level in resist development. A few U.S. companies are investing "in house" or in collaboration with resist vendors. It will be interesting to see how this conflict between the demand for small volumes of highly sophisticated specialty polymers and the high cost of developing such materials plays out over the next few years.

Compact Disk Technology

Compact disks have emerged as the dominant recording medium for the musical entertainment field. Information is recorded as a series of pits on radial or concentric tracks that extend from the inner to the outer diameter of the disk. The pits are typically 0.25 μm deep, 0.5 μm wide, and up to 3.5 μm long. The information is read by means of a laser beam that is reflected when it falls on the flat of the disk but is almost entirely deflected when it falls on a pit. This digital bit stream is converted to an analog signal to reproduce the music. The disk itself is made of a polymer by means of a process that is technically demanding and economical.

The manufacturing sequence consists of encoding the pit pattern onto a glass master. This is accomplished by means of a lithographic process employing a polymer resist and an irradiating laser. The open areas thus formed are etched to form the pits. Nickel is then vacuum deposited, thickened, and formed into a negative "stamper." The stamper is then seated into a mold cavity, and CDs are produced by injection molding of polycarbonate or poly(methyl methacrylate). The molding process is carried out at very high pressures, and dust particles must be avoided. A class-10 cleanroom is usually required to achieve quality replication and long stamper life.

Molded-in stress is a major consideration. When polycarbonate flows in the submillimeter thickness of optical disks for several centimeters, there is significant molecular orientation that manifests itself as birefringence or optical distortion. Material and process parameters have been refined to control birefringence and maintain replication integrity. New long flow grades of the materials have been developed specifically for the CD market.

Polymeric Materials for Photonics

Photonics is a technology analogous to electronics in which the photon replaces the electron as the working particle. Many of the applications now accomplished electronically, including transmission, switching, amplification, and modulation, can also be realized using photonics, and there are advantages to be gained by converting to a photon-based technology in some areas. Transmission of light in fiber-optic systems is the direct analogy of electrical transmission in coaxial cable systems.

Fiber-optic systems are now in place all over the world, and they handle much of the world's long-distance telephone traffic. The transmission medium of the fibers employed is based on inorganic glasses, but polymers are used for protective coatings and in cabling structures. Polymers can also be made into optical fibers, but the loss is considerably larger than with the inorganic fibers and only short-distance applications are realistic. The main advantage of polymer fibers is their flexibility when made in larger diameters, which are easier to splice. Today, fiber-optic cables are generally terminated at the area substation level, where the optical signal is converted back to an electrical signal for transmission to the customer. This conversion process is necessary because the optical components needed to reach the individual telephone or terminal are not available at sufficiently low cost at this time. What is required to allow fiber to be connected to the home are inexpensive optical switches and amplifiers, which will enable the advantages of broad-band communications to be brought to every subscriber. Polymeric organic materials will play a major role in the realization of optical technology as fiber to the home becomes a reality.

Two kinds of optical technology need to be developed and commercialized before the photonics revolution can be fully realized. The first is linear optical technology, which includes not only the long-distance fibers mentioned above, but also shorter fibers and the optical equivalent of printed wiring boards of the electrical domain. These optical circuits can be created today by means of a photolithographic procedure in which lines of high refractive index are formed in thin polymer films by photochemical techniques. The circuit pattern is defined by irradiating a photoresist through a mask. The substrate film bared by development of the resist is then exposed to light, which causes the chemistry, for example, the polymerization of monomers, that gives rise to the increase in refractive index needed to form an optical guide.

Nonlinear optical materials will also be required for the manufacture of switches, modulators, and amplifiers, and this technology has not progressed as far as the linear domain. Demonstration-of-principle devices have been fabricated that prove the viability of polymers in this application. The necessary switches, amplifiers, and modulators can be made today with inorganic materials, but there is some question whether these realizations can be combined and manufactured in large volume at sufficiently low cost. Organic thin film technologies may fit the economic as well as the technological requirements, but many advances will be required and the outcome is uncertain.

A nonlinear polymer in general has two components: the polymer itself and an optically nonlinear molecule (a chromophore) that is either chemically attached to the polymer or dissolved in it. In order for the polymer-chromophore system to be optically nonlinear, the chromophores must be aligned such that on average they are all pointing in the same direction within the polymer matrix. This alignment is accomplished through a process called poling. The polymer is poled by cooling it through the glass transition temperature while it is in a very strong electric field, and the order induced by the field is frozen in.

Poled polymeric systems have process and property advantages over their inorganic crystalline competitors. The polymers can be formed into thin films and lithographically patterned, and they can be chemically modified to tailor and improve bulk properties. There are disadvantages as well, in that the orientation in the poled polymer systems tends to decay with time, a problem that can probably be overcome.

Polymeric Light-emitting Diodes

Recently, light-emitting diodes (LEDs) based on conducting polymers have been achieved in a number of laboratories around the world. The active element is a thin film structure based on a modified poly(phenylene vinylene) (PPV), with a metal film as the electron injector and polyaniline as the hole injector. Various colors have been demonstrated, and the operating characteristics are competitive with inorganic LEDs. Highly flexible devices have been fabricated supported on a poly(ethylene terephthalate) base. The possibility of making large-area displays exists.

Much research and development remains to be done. For example, low-work-function metals are required, and they are difficult to passivate. However, the simplicity of fabrication of the laboratory devices, involving spin casting from solution, is promising if the problem of limited device lifetime can be solved.

Polymers for Electrophotography

One of the major applications of polymers with tailored electronic and optical properties has been in electrophotography for copier, duplicator, and printer applications. In this application an electroactive polymer is used as one component of the light-sensitive element used for creating the latent electrostatic image of an original subject. The image source can be light reflected from a document and focused onto the surface of the photoreceptor or a digital file of an original image, which is used to control a laser beam that is scanned over the surface of the photoreceptor. The electrostatic image is rendered visible by dusting the surface of the photoreceptor with an electrostatic powder composed of a pigment-loaded thermoplastic polymer. The latent image can then be transferred to paper by a combination of pressure and electrical bias and then fused to the paper by heating.

The photoreceptor itself was the key invention that enabled the development of electrophotography as a commercial success. The original photoreceptor materials were based on selenium and its alloys as well as group II-VI and other semiconductor materials. Because of the poor mechanical properties of selenium and its alloys, photoreceptors had to be fabricated on rigid metallic drums. This, in turn, dictated relatively cumbersome and expensive copier machine architectures. These materials had a number of shortcomings, including degradation of photoconductive properties, instabilities in surface properties leading to incomplete toner transfer, and catastrophic abrasion.

Research efforts in several industrial and university research laboratories were successful in identifying polymeric materials that exhibited photoconductivity. The early photoconductive polymers were mainly sensitive to UV light. The copying process, however, requires differential reflectivity from the printed areas of the original document, which is very low for UV light but much higher for visible light sources. The need for visible light sensitivity was therefore apparent. Eventually, the problems associated with spectral sensitivity and a variety of other technological requirements were solved, and it was clear that the polymeric materials could be used advantageously in copier and printer technology.

The latent electrostatic image is formed by first depositing a layer of ionic charge from a corona discharge onto the photoreceptor surface. This induces an equal but opposite charge on the metal layer below, resulting in the formation of an electric field within the photoconductor layers. As light passes through the transport layer and is absorbed by the photosensitive pigment layer, the pigment molecules are photoionized with the assistance of the internal electric field to form mobile charge carriers. The negative photogenerated charge in the film drifts under the influence of the electric field to the metal, and the positive charge drifts through the transport layer and neutralizes some of the ionic charge that was deposited on the surface. Since photogeneration will occur only where light strikes the photoreceptor, a pattern of ionic charge corresponding to the original image is formed on the surface. The surface potential associated with this charge distribution is used to attract the toner as described above.

Photoreceptor belts have been engineered to exhibit excellent mechanical properties, and this achievement has allowed the design of compact and cost-effective copier and printer architectures. Useful lifetimes, photosensitivities, and mechanical durability of photoreceptors have been extended well beyond those of the original selenium-based drums. Photoreceptor wavelength sensitivities have now been extended to the near-infrared so that inexpensive diode lasers and light-emitting diode arrays can be used for digital printing applications.

Polymers in Holography

A light interference pattern comprising relatively large light intensity variations on a microscopic scale is created where two previously separate light beams from the same laser intersect. A hologram is a physical record of such a pattern and is formed by exposing a photosensitive recording film to the interference pattern. When a hologram is illuminated with one of the two laser beams used in its recording, it produces a light beam that is essentially identical to the other recording beam. Familiar image holograms are usually produced from a simple collimated or diverging light beam, called a reference beam, and a beam formed by scattering light from a complex three-dimensional solid object. When such a hologram is illuminated with the reference beam, it produces a light beam that appears to come from the solid object used in its recording.

Holograms can also be made from light beams produced by conventional optical elements such as lenses and mirrors. The resulting holograms, called holographic optical elements (HOEs), perform optical functions of the elements used in their recording. One HOE type, for example, is recorded with a collimated light beam and a light beam that converges to a focal point. When illuminated with the collimated beam, the resulting hologram will produce a focused beam; it acts, therefore, as a holographic lens. HOEs have important advantages over conventional optical elements. They are lightweight and compact and can take the place of heavy and bulky glass elements. They can be made very large or very small. They can replace expensive conventional optics for the production of arbitrarily complex light beams. They can be inexpensively mass produced.

There is a wide variety of current and potential applications for holography. The use of holographic three-dimensional images is probably the most familiar application. These images are typically used on credit cards and for product advertisement and promotion. In these applications, holograms add both eye appeal and security. Holographic images are also used in nondestructive testing. Holographic optical elements can be made in large thin films for use in solar lighting control and solar energy collection, and they can be made very small for use in optical communication systems. Narrow-band holographic mirrors may also be useful for laser eye protection. Optical computing, pattern recognition, and very-high-density information storage are other potential applications of holography.

Many holographic applications require the high performance that is possible only with "phase" holograms, in which the original interference pattern is recorded as a refractive index variation. Conventional phase hologram recording materials have, unfortunately, limitations that have inhibited the growth of practical holography. Phase hologram recording based on photopolymerization is a relatively recent development, and it promises to overcome important problems of current recording materials.

Holographic photopolymer systems comprise, as major components, a film-forming polymer (often called the binder), a photoinitiation system, and a monomer. The polymer binder aids in coating the appropriate substrate and helps to maintain film integrity during holographic exposure and subsequent processing. Properties of the binder can also strongly influence both the shelf life of the coated film and the rates and extents of photochemical reactions that occur during hologram formation.

Monomers join in a chain reaction during laser exposure. In fact, the relatively good light sensitivity of photopolymers results from the large number (100 to 1,000) of monomer units that react per absorbed photon. The chemical and physical changes associated with monomer polymerization preserve the interference pattern created during exposure as a corresponding pattern of refractive index variation. Numerous and diverse chemical and physical requirements greatly limit monomer choice. Shelf life and light sensitivity must be balanced. Film clarity and image stability are essential. Large refractive index changes are desirable. The monomer must also be compatible with the other components of the system.

The ideal material does not yet exist. There is an excellent chance, however, that continued research with photopolymer systems will produce new holographic recording materials that will make practical many potential applications of holography.

Conclusions

The foregoing examples illustrate the breadth of application of advanced polymeric materials in applications that are not generally recognized. In these applications, the polymer is in some sense the active element that plays the central role. No other class of materials can rival its range of properties, flexibility in processing, and potential for low cost. Quality of performance is an essential and challenging feature that is being demonstrated by polymeric materials in an impressive array of applications. And the polymer revolution in this arena is just beginning.

The importance of polymers in advanced technology is a key factor in the future of materials development, as indicated in the following applications:

- Polymer dielectrics in electronics offer the basis for the smallest circuits and the highest speed of operation.

- Conducting polymers have been commercialized in rechargeable batteries and offer the greatest promise for high energy storage with low weight.
- Polymer sensors exist for chemical species, thermal and acoustic radiation, temperature, pressure, humidity, ionizing radiation, electric charge, and more.
- Buildings can be equipped with a network of optical fibers linking remote locations with a management console. The polymer sensors can be built into the optical fibers to report the presence of toxic gases or to turn off unneeded lights to conserve energy.
- Implanted sensors can detect the glucose level in blood and call for insulin injections by means of an implanted pump, as needed.
- Electromagnetic shielding will become increasingly necessary, and conducting polymers offer solutions that are conveniently fabricated in complex shapes.
- Polymer resists are the basis for the microlithography that makes integrated circuit electronics possible. They are also the basis for the emerging field of micromechanics, which could produce machines smaller than a human cell.
- High-density information storage is available through compact disk technology, and improved polymers will improve the performance of this medium. In the future, polymer-based holographic devices could revolutionize the storage and manipulation of information.
- Polymers offer solutions to critical economic problems facing the introduction of photonics, the light analog of electronics. The couplers, splitters, and other elements of photonic "circuit boards" all admit to polymeric solutions that may provide the economic breakthrough needed for the photonic revolution. Broad-band communications can be brought directly to the home and office by polymer or glass fibers, using polymeric photonic circuits.
- "Smart" windows based on polymeric materials could reflect light when the sun is too bright and transmit light when it is not.
- The fabrication of liquid crystal display devices for computers and television can be facilitated and the robustness of the product enhanced by the incorporation of conductive, transparent polymer films.
- Light-emitting diodes based on flexible polymeric films have been fabricated and are likely to find diverse applications in the future.
- Electrophotography is now based on polymeric photoactive materials, and these have made possible many improvements, such as compact and convenient machine architecture, durability of machines, and long-term print quality.
- Polymers are now the recording medium of choice for holography in many applications. This technology offers the promise of ultrahigh-density information storage.

The field is flourishing, and the future is bright. The United States must participate vigorously in this emerging area, from research to development to commercialization. Competition in the field is worldwide and moving rapidly ahead.

REFERENCES

Physicians' Desk Reference. 1994. 48th ed. Montvale, N.J.: Medical Economics Data.
Sze, Simon. 1988. *VLSI Technology*. 2nd ed. New York: McGraw-Hill Book Co.

3

Manufacturing: Materials and Processing

Materials as a field is most commonly represented by ceramics, metals, and polymers. While noted improvements have taken place in the area of ceramics and metals, it is the field of polymers that has experienced an explosion in progress. Polymers have gone from being cheap substitutes for natural products to providing high-quality options for a wide variety of applications. Further advances and breakthroughs supporting the economy can be expected in the coming years.

Polymers are derived from petroleum, and their low cost has its roots in the abundance of the feedstock, in the ingenuity of the chemical engineers who devised the processes of manufacture, and in the economies of scale that have come with increased usage. Less than 5 percent of the petroleum barrel is used for polymers, and thus petroleum is likely to remain as the principal raw material for the indefinite future. Polymers constitute a high-value-added part of the petroleum customer base and have led to increasing international competition in the manufacture of commodity materials as well as engineering thermoplastics and specialty polymers.

Polymers are now produced in great quantity and variety. Polymers are used as film packaging, solid molded forms for automobile body parts and TV cabinets, composites for golf clubs and aircraft parts (airframe as well as interior), foams for coffee cups and refrigerator insulation, fibers for clothing and carpets, adhesives for attaching anything to anything, rubber for tires and tubing, paints and other coatings to beautify and prolong the life of other materials, and a myriad of other uses. It would be impossible to conceive of our modern world without the ubiquitous presence of polymeric materials. Polymers have become an integral part of our society, serving sophisticated functions that improve the quality of our life.

The unique and valuable properties of polymers have their origins in the molecular composition of their long chains and in the processing that is performed in producing products. Both composition (including chemical makeup, molecular size, branching and cross-linking) and processing (affected by flow and orientation) are critical to the estimated properties of the final product. This chapter considers the various classes of polymeric materials and the technical factors that contribute to their usefulness. In spite of the impressive advances that have been made in recent years, there are still opportunities for further progress, and failure to participate in this development will consign the United States to second-class status as a nation.

MATERIALS

Structural Polymers

The familiar categories of materials called plastics, fibers, rubbers, and adhesives consist of a diverse array of synthetic and natural polymers. It is useful to consider these types of materials together under the general rubric of structural polymers because macroscopic mechanical behavior is at least a part of their function. Compared with classical structural materials like metals, the present usage represents a considerable broadening of the term. As shown in Table 3.1, man-made plastics, fibers, and rubber accounted for U.S. production of about 71 billion pounds in 1992 (*Chemical & Engineering News*, 1993), and production has tripled over the last 20 years. The price received by the original manufacturer ranges from roughly $0.50 to several dollars per pound, depending on the material. At $20 per barrel, crude oil costs about $0.06 per pound, and so conversion to polymers represents considerable value added. Because these materials go through several manufacturing steps before reaching the final consumer, the ultimate impact on the national economy is measured in the hundreds of billions of dollars each year.

TABLE 3.1 U.S. Production of Some Man-Made Structural Polymers, 1992

	Pounds (billions)
Plastics	57.6
Fibers	9.1
Rubber	4.2

SOURCE: Data from *Chemical & Engineering News* (1993), p. 44.

These materials have many different chemical and physical forms, such as cross-linked versus non-cross-linked, crystalline versus amorphous, and rubbery versus glassy. More recently, structural polymers having liquid crystalline order have become important. Structural polymers are rarely used in the pure form but often contain additives in small quantities, such as antioxidants, stabilizers, lubricants, processing aids, nucleating agents, colorants, and antistatic agents or, in larger quantities, plasticizers or fillers. There is rapid growth in the areas of blends and composites. In composites, a material of fixed shape, such as particles (filler) or fibers, is dispersed in a polymer matrix. The filler or fiber may be an inorganic material or another organic polymer. Blends (or alloys) on the other hand consist of two or more polymers mixed together and differ from composites in that the geometry of the phases is not predetermined prior to processing. Some polymers are used for many different purposes. A good example is poly(ethylene terephthalate), or PET, which was originally developed as a textile fiber. It is now used in film and tape (virtually all magnetic recording tape is based on PET), as a molding material, and as the matrix for glass-filled composites. One of its largest uses is for making bottles, especially for soft drinks. PET is also used in blends with other polymers, such as polycarbonate.

Plastics

The word "plastic" is frequently used loosely as a synonym for "polymer," but the meaning of "polymer" is based on molecular size while "plastic" is defined in terms of deformability. Plastics are polymeric materials that are formed into a variety of three-dimensional shapes, often by molding or melt extrusion processes. They retain their shape when the deforming forces are removed, unlike some other polymers such as the elastomers, which return to their original shape. Plastics are usually categorized as thermoplastics or thermosets, depending on their thermal processing behavior.

Thermoplastics Thermoplastics are polymers that soften and flow upon heating and become hard again when cooled. This cycle can be repeated many times, which makes reprocessing during manufacturing or recycling after consumer use possible using heat fabrication techniques such as extrusion or molding. The polymer chains in thermoplastics are linear or branched and do not become cross-linked as in the case of thermosets. While there are many different chemical types of thermoplastics, those made from only four monomers (ethylene, propylene, styrene, and vinyl chloride) account for about 90 percent of all thermoplastics produced in the United States (Figure 3.1). Of these four types, polypropylene has grown most rapidly in recent years—production has increased eightfold over the past two decades. Thermoplastic polyesters, primarily PET, are growing even more rapidly at the present time (driven mainly by

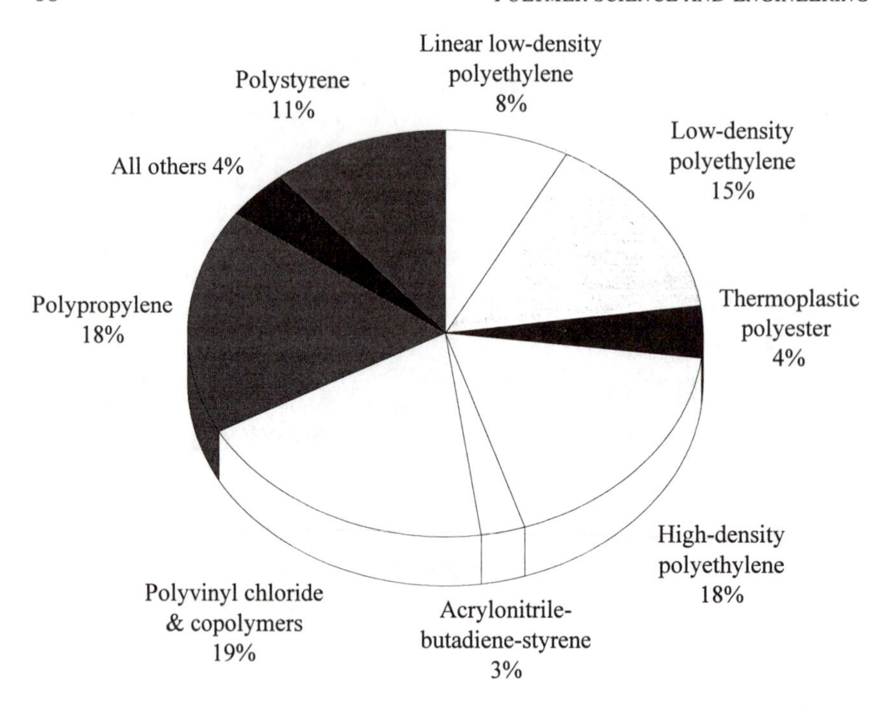

FIGURE 3.1 U.S. production of thermoplastics by type, 1990. SOURCE: Reprinted with permission from *Chemical & Engineering News* (1991), p. 54. Copyright© 1991 by the American Chemical Society.

packaging applications), with current sales nearly one-quarter of those for polypropylene. For the long term, the majority of commodity thermoplastics are expected to follow their traditional growth (*Chemical & Engineering News*, 1992), with continued opportunities for both process and product innovation. Future activities will focus strongly on recycling. In the case of PET, recycling can be accomplished by chemical depolymerization to monomers or oligomers followed by repolymerization to PET or other products. Such processes are currently in use for products that come into contact with food, while simple reprocessing is used for less critical products.

The so-called engineering thermoplastics, which include the higher-performance, more expensive polymers such as the polyacetals, polycarbonates, nylons, polyesters, polysulfones, polyetherimides, some acrylonitrile butadiene styrene (ABS) materials, and so on, have generally exhibited stronger growth than the commodity plastics (see Table 3.2). These materials generally have greater heat resistance and better mechanical properties than the less expensive commodity thermoplastics and, therefore, are used in more demanding applications, such as aircraft, automobiles, and appliances. A major area of development is

TABLE 3.2 Pounds of Selected High-volume Engineering Thermoplastics Sold in the United States, 1981 and 1991

	Pounds (millions)		Percentage Increase
	1981	1991	Since 1981
Thermoplastic polyesters	1,230	2,550	107
Acrylonitrile butadiene styrene	968	1,130	17
Nylon	286	556	94
Polycarbonate	242	601	148
Poly(phenylene oxide)-based alloys	132	195	48
Polyacetal	88	140	59

SOURCE: Data from *Modern Plastics* (1982, 1992).

new blends or alloys of engineering plastics that are designed for specialty market products and are usually quite tough and chemical resistant. (The area of blends and alloys is reviewed separately below.) New products and advances in processes have resulted from the ring-opening polymerization of cyclic oligomers; for example, new developments in polycarbonates are particularly noteworthy. Other new products can be expected based on copolymers, and entirely new polymers are under development.

A further category sometimes referred to as high-performance engineering thermoplastics commands even higher prices for yet higher levels of performance. These include highly aromatic polymers such as poly(phenylene sulfide), several new polyamides, polysulfones, and polyetherketones. Development of new molecular structures has dominated this sector. Polymer chains with quite rigid backbones have liquid crystalline order, which offers unique structural properties as described below.

Figure 3.2 shows the major categories of use for thermoplastics. Approximately one-third are used in packaging, primarily containers and film. The data in Figure 3.2 are dominated by the huge volume of the five or so commodity thermoplastics; hence, the products with greater value based on engineering or advanced thermoplastics do not emerge in true proportion to their contribution to the national economy.

To understand the diversity of products and opportunities that is possible, it is useful to review developments that have occurred in thermoplastics based on ethylene, one of the simplest monomers possible. Commercial production of polyethylene commenced in England during the early 1940s using a free radical process operating at very high pressures (30,000 to 50,000 psi). The structure proved to be far more complex than the simple textbook repeat unit, $-CH_2CH_2-$, would suggest (Figure 3.3). The backbone has short- and long-chain branches. The short-chain branches, typically four carbons long, interfere with the ability

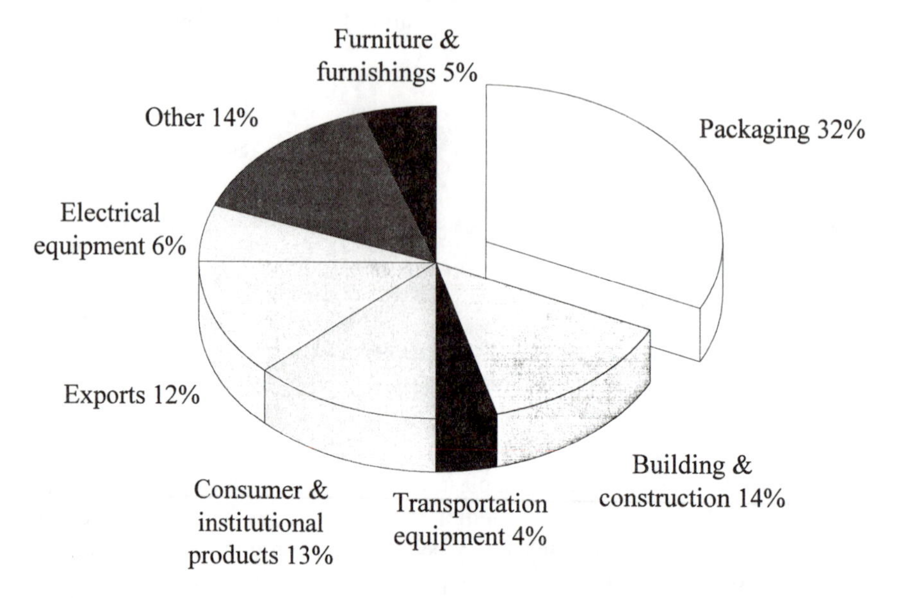

FIGURE 3.2 Categories of uses for thermoplastics in the United States, 1990. SOURCE: Reprinted with permission from *Chemical & Engineering News* (1991), p. 56. Copyright© 1991 by the American Chemical Society.

of the chain to crystallize, thus affecting solid-state properties, while the long-chain branches (comparable in length to the backbone itself) mainly affect melt rheological or flow properties that influence processing behavior. Because the short-chain branches reduce crystallinity and, thus, density, this material is called low-density polyethylene (LDPE). In the late 1950s, a linear or unbranched form of polyethylene was developed as a result of advances in coordination polymerization catalysis. An accidental finding by K. Ziegler in the early 1950s at the Max Planck Institute of Mulheim, Germany, resulted in a fundamentally new approach to polyolefins. It was found that transition metal complexes could catalyze the polymerization of ethylene under mild conditions to produce linear chains with more controlled structures. As a result, this polymer was more crystalline with higher density, and it became known as high-density polyethylene (HDPE). Similar catalytic procedures were used by G. Natta to produce crystalline polypropylene. The properties of this polymer are a result of unprecedented control of the stereochemistry of polymerization.

Because of the effects of molecular structure on crystallinity, HDPE is as much as 5 times stronger and 1 order of magnitude stiffer than LDPE. The newer material did not replace the older one; it was used for different purposes. In the 1970s, the high-pressure LDPE process became increasingly expensive relative to the lower-pressure HDPE process. The cost factor plus innovations in

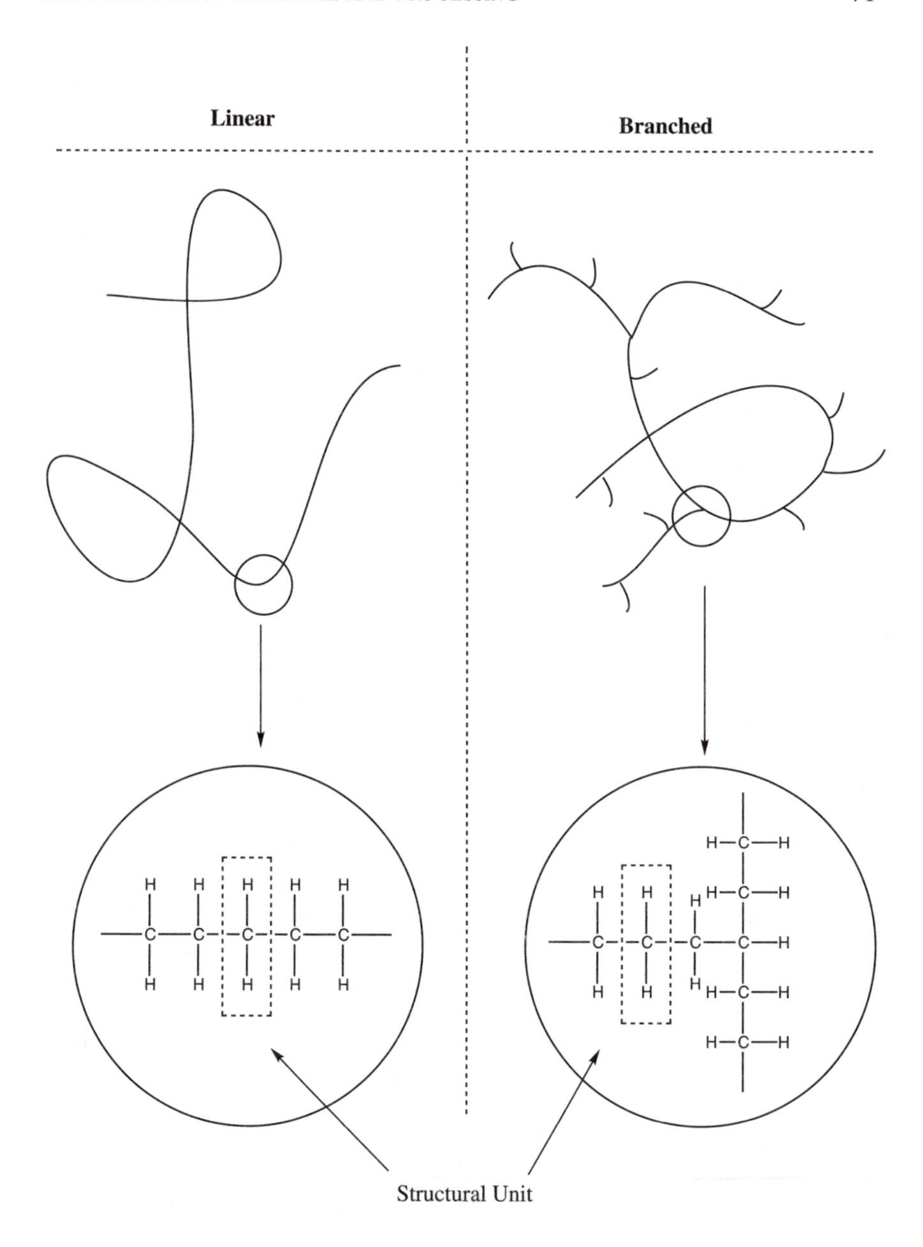

FIGURE 3.3 Schematic of the structure of high-density polyethylene, low-density polyethylene, and linear low-density polyethylene.

catalysts and process technology led to a new material that had most of the attributes of LDPE but was produced by a more economical low-pressure process similar to that used for HDPE. It is a copolymer of ethylene and an alpha-olefin (like butene-1, hexene-1). Thus, short-chain branches of controlled length and number are introduced into the chain without any long-chain branches, and the material is called linear low-density polyethylene (LLDPE; see Figure 3.3). Production of this material grew at a rate of about 20 percent per year during the 1980s to current usage of about 5×10^9 pounds per year. As a result, the production of LDPE initially declined, but its production has been growing again since 1986. Construction of new high-pressure production facilities may be required in the next decade to meet demands. Currently this is the only process by which copolymers can be made with polar monomers such as vinyl acetate or acrylic acid. HDPE is fabricated primarily by molding. Blow-molded food bottles and auto gasoline tanks constitute major markets. Very large containers made by rotational molding represent a specialized growth area. A process known as "gel spinning" has been commercialized, which produces fibers of ultrahigh-molecular-weight polyethylene. The less crystalline LDPE and LLDPE are primarily extruded into film products, with each having specialized uses. New technology based on single-site metallocenes holds promise for the production of a new range of products.

This brief review of the history and future prospects for olefin polymers illustrates the need for research of all types (e.g., catalysis, process, and structural characterization) in order to capitalize on economic opportunities. These materials are complex in terms of molecular structure, and so there are many ways to tailor their behavior provided the basic knowledge and tools for structural determination are available and are integrated with innovative process technology. Much of the present research is directed toward the design of catalysts that yield materials that are easier to process. Rapid progress has resulted from an integration of catalyst synthesis and reactor and process design. As a recent example, a new polyolefin alloy product has been developed by exposing a designed catalyst to a series of different olefin monomer feeds to produce a polymer particle that is composed of polymers with different properties. Extrusion of those particles results directly in a polymer alloy.

Structural thermoplastics are a vital part of the national economy, and considerable opportunity remains for economic growth and scientific inquiry. New specialized materials will continue to offer rewards in the marketplace. At the high-performance end, several entirely new polymer structures are likely to emerge over the next decade. A major part of the growth in "new" materials will be in the area of blends or alloys. The vitality of thermoplastics cannot be judged only on the basis of the introduction of what might be called "new materials." Continuous improvement and diversification of existing polymers consitute another measure. One source estimates that the number of "grades" of existing polymers tripled during the 1980s (*Chemical & Engineering News*,

1991). This trend is expected to continue but will require greater sophistication in terms of process technology, characterization, and structure-property relationships (especially modeling) than has been required in the past.

Thermosets Thermoset materials are broadly defined as three-dimensional, chemically resistant networks, which in various technologies are referred to as gels, vulcanizates, or "cured" materials. Applications as diverse as coatings, contact lenses, and epoxy adhesives can be cited. Thermosets are defined here as rigid network materials, that is, as materials below their glass transition temperature. Thermosets are formed when polyfunctional reactants generate three-dimensional network structures via the progression of linear growth, branching, gelation, and postgelation reactions. The starting monomers must include at least some reactive functionality greater than two, which will ensure that as the reaction proceeds, the number of chain ends will increase. They will eventually interconnect to produce a gelled network material. This process may be followed by observing the viscosity increase as a function of time or from the percent reaction completed. In many cases, this can be predicted mathematically. As the gel begins to form, the soluble fraction decreases and eventually is eliminated altogether.

An important consideration with respect to rigid thermosetting networks is the extensively studied interrelationship between reactivity, gelation, and vitrification. As the reaction proceeds, the glass transition temperature rises to meet the reaction temperature, and the system vitrifies; that is, the motion of the main chain stops. At this point, the reaction essentially stops for all practical purposes. This has been conveniently described in terms of a time-temperature-transformation cure diagram. Thermosetting systems can be formed either by chain or step polymerization reactions. The chemistry of thermoset materials is even now only partially understood, because they become difficult to characterize once they reach the three-dimensional insoluble network stage. Thermal and dynamic mechanical methods have been widely used to characterize these materials, and solid-state nuclear magnetic resonance (NMR) has begun to have some impact on this problem.

Thermoset materials make up approximately 15 percent of the plastics produced in the United States. Figure 3.4 shows recent data on the production of the various types of thermosets and their uses. Phenolics make up the largest class of thermoset materials. Some polyurethanes are classified as thermosets, although many urethane and urea materials can be produced in linear thermoplastic or soluble forms, such as the well-known elastomeric spandex fibers. Urea-formaldehyde-based materials continue to be significant and, in fact, were the systems used in the first "carbonless" paper. Unsaturated polyesters are derived from maleic anhydride and propylene glycol, which are then dissolved in styrene and cross-linked into a network. They have gained significant importance in

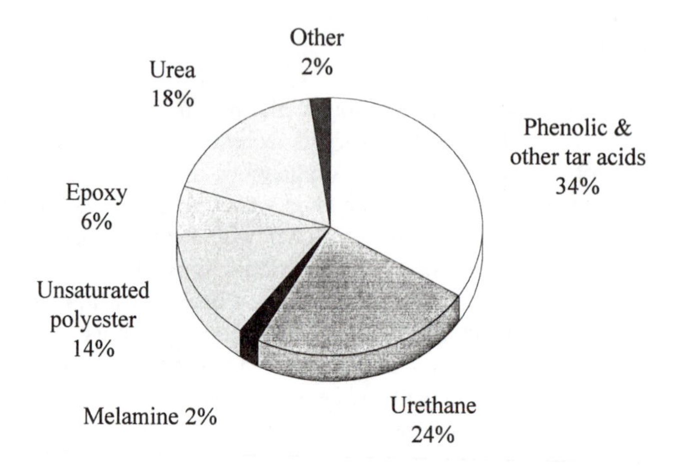

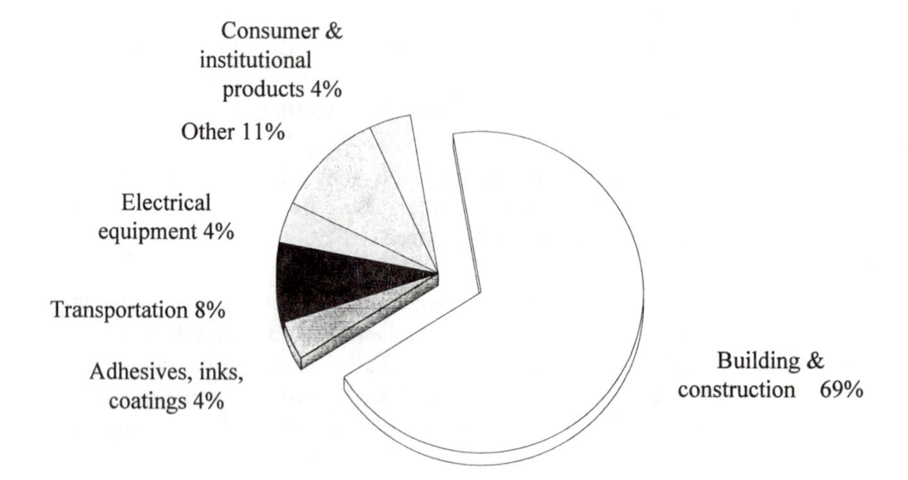

FIGURE 3.4 U.S. production of thermosets by type for 1990 (top) and their areas of use in 1989 (bottom). SOURCE: Reprinted with permission from *Chemical & Engineering News* (1991), p. 39. Copyright© 1991 by the American Chemical Society.

automobiles and construction. The resulting glass-reinforced composites are frequently called sheet molding compounds (SMC).

Thermoset materials, although smaller in total volume than the thermoplastics, are used in a number of very high performance applications, such as matrix resins or structural adhesives in composite systems such as those used for aerospace applications. These composites are normally reinforced with glass, aramid, or carbon fibers. Important examples of such matrix materials include the epoxies, bismaleimides, cyanates, acetylenes, and more recently, benzocyclobutene systems. The existing database for matrix resins and structural adhesives is much more established for thermosets than it is for high-performance thermoplastics such as the poly(arylene ether ketones), certain polyaryl imides, and poly(phenylene sulfide). Major research needs in the area of polymer-based composites include better ways to improve the toughness of thermosetting systems and better techniques for processing those formed from high-performance thermoplastics.

Advances in processing and toughening thermosets are occurring on several fronts. Methods for generating the network have been investigated by many organizations. The most conventional methods involve use of a thermal-convection-oven-type curing, often in autoclaves. However, recently there has been considerable effort in electromagnetic (or microwave) processing of high-performance polymeric matrix resins, particularly for structural adhesives and composite structures. An approach for "toughening" that has been investigated over the last 10 years involves the incorporation of either rubbers or reactive engineering thermoplastics into networks, such as epoxies, to develop a complex morphology. Here the added material is dispersed as isolated domains or forms co-continuous morphologies. Most of the original studies focus on rubber toughening, and an extensive body of literature deals with utilization of carboxyl functional nitrile rubbers to toughen epoxy adhesives. More recently, advantages associated with the utilization of engineering thermoplastics have been realized. These include, for example, the ability to retain stiffness and thermooxidative stability, as well as in some cases, chemical resistance. These properties are often severely diminished with rubber-toughened thermosetting systems. Fracture toughness can be significantly improved. This is significant in terms of improving the durability of advanced organic materials utilized in structural adhesives and composites. The interfacial adhesion between the separate polymer phases, as well as their proportions, morphology, and molecular characteristics, is of prime significance in improving fracture toughness.

Other forefront areas include the development of new chemistries and, in particular, better characterization of leading candidate materials. The bismaleimides are considered to be somewhat more thermally stable than the epoxy materials and are being seriously considered for various applications, such as the high-speed civil transport airplane, which is planned for commercialization in the next 10 years. Aspects of the flammability of these materials are also crucial.

Aryl-phosphine-oxide-containing materials show considerable promise for producing advanced organic materials with significantly improved flammability resistance. A new development is the possibility of bridging organic and inorganic materials to produce organic-inorganic composite networks.

Elastomers

Elastomers, or rubbers, are soft and compliant polymers that are able to experience large, reversible deformations. Only long-chain polymers are capable of this type of elasticity. Elastomers are typically amorphous, network polymers with lower cross-link density than thermoset plastics. Most thermosets can be made to function as elastomers above their glass transition temperatures.

Historically, elastomers have played an important role in the industrialization, prosperity, and security of the United States. Synthetic elastomers were born of necessity during World War II, when the United States was cut off from most of its supplies of natural rubber in Southeast Asia. Low-temperature emulsion polymerizations were developed to produce highly successful synthetic rubbers, particularly styrene-butadiene copolymers. In one of the most remarkable success stories in modern industry, a production capacity of 1.5 billion pounds per year was reached in 1945. This industry continues in the United States (see Table 3.3 for production figures for various types of synthetic rubber) and in other industrialized countries. Annual production figures have declined in recent years owing to a number of factors, including advances in the use-life of tires.

Originally, all elastomers were thermosets or chemically cross-linked materials, and so their flexibility in processing, especially reprocessing or recycling, was severely limited. Thermoplastic elastomers represent a current major growth area that comprises a growing number of chemical concepts. The first materials were styrene-based block copolymers that phase separate at the molecular level to produce relatively hard polystyrene domains, which act as temporary, physi-

TABLE 3.3 U.S. Production of Synthetic Rubber, 1992

	Pounds (billions)
Ethylene-propylene	0.45
Nitrile	0.16
Polybutadiene	1.02
Styrene-butadiene	1.75
Other[a]	0.86

SOURCE: Date from *Chemical & Engineering News* (1993).

[a]Includes, for example, butyl, polychloroprene, polyisoprene, silicone, and other synthetic elastomers.

cal cross-links. The resulting elastomer is thermoplastic, and it is possible to reprocess it by simply heating it to above the glass transition temperature of polystyrene. It is thus a reprocessible elastomer. These materials are the result of the development of anionic polymerization methods, which are now practiced on a large scale in spite of the tremendous experimental difficulties associated with the organometallic initiators used in this process. Similar concepts have been implemented commercially for polyurethanes, polyesters, polyether-amides, and so on. Other versions are the so-called dynamically vulcanized blends of plastic and rubber that can be molded or extruded like thermoplastics. Current characterization techniques do not permit probing all of the potentially critical structural issues of such complex materials.

With regard to theory, there is a need for a better understanding of the topology of the network structure that is required for the recoverability exhibited by elastomers. More specifically, we need to know how to characterize entanglements and their effects on mechanical properties. Such topological features would be expected to have large effects on both equilibrium and dynamic properties, and their control could help greatly in the design of more competitive elastomers. Although the deepest insights into rubberlike elasticity will almost certainly come from molecular theories, phenomenological approaches are also frequently useful, particularly for practical purposes. These theories attempt to fit stress-strain data using a minimal number of parameters, which are then used to predict other mechanical properties. There is also a need for more experimental data on deformations other than simple elongation and swelling, which (because of their simplicity) are the ones used in the overwhelming majority of elasticity studies. One benefit would be additional, discriminating data for evaluating elasticity theories. Another would be the better understanding of the properties of elastomers for conditions under which they are frequently used. An understanding of segmental orientation of chains in deformed networks is essential for an understanding of strain-induced crystallization. Such crystallization greatly enhances the mechanical properties of an elastomer, and its control could be of considerable competitive advantage. Advances in theory, as well as additional experiments, are required for progress in this area.

There is increasing interest in the study of elastomers that also exhibit mesomorphic behavior, from liquid crystalline entities either in the chain backbone or in the side chains. These materials combine some of the most intriguing properties of liquid crystalline molecules of low molecular weight with the elastomeric properties of polymeric networks. Materials with this unique behavior should be exploited. An example is the orientation of an anisotropic phase by a mechanical force, in analogy to the use of electric or magnetic fields on low-molecular-weight mesogens.

A new subject in the area of rubberlike elasticity is the phenomenon of gel collapse, in which a swollen network abruptly deswells (shrinks) in response to a relatively small change in its environment. The collapse can be triggered by changes in temperature, pH, ionic strength, or solvent composition. In films and fibers the collapse is rapid enough for these systems to have potential applications as mechanical switches, artificial muscle, mechanochemical engines, and so on. Exploiting this new development will require advances in theory, as well as additional experiments in actual functioning devices.

A particularly challenging problem is the development of a more quantitative molecular understanding of the effects of filler particles, in particular carbon black in natural rubber and silica in siloxane polymers. Such fillers provide tremendous reinforcement in elastomers in general, and how they do this is still poorly comprehended. Certainly the bonding between the reinforcing phase and the elastomeric matrix is critical. Investigation of the bonding between the biopolymer elastin and the collagen fibers that are threaded through it for reinforcement could provide valuable insights into this problem. A related problem exists in the hybrid organic-inorganic composites.

Finally, there is a need for more high-performance elastomers, which remain elastomeric to very low temperatures but are relatively stable at very high temperatures and resist hostile environments. The elastomeric ethylene-propylene-diene monomer (EPDM) rubbers, made by copolymerization of ethylene, propylene, and a diene using Ziegler catalysts, are particularly resistant to ozone. The polysiloxanes are one of the most important classes of high-performance elastomers and are being developed and improved in most industrialized countries. The fluoroelastomers are another class that is under intense development. Polyphosphazenes have rather low glass transition temperatures in spite of the fact that the skeletal bonds of the chains are thought to have some double-bond character. There are, thus, a number of interesting problems related to the elastomeric behavior of these unusual semi-inorganic polymers.

Fibers

A fiber may be defined as a structure whose length is much greater than its cross-sectional dimension. The diameter of fibers is characterized in dTex, a unit of linear density corresponding to the weight in grams of a 10,000-meter length of the fiber (the same units are used to describe individual filaments and multifilament yarns). Typical filament dTex values run from 1 to 10. One gram of a 1-dTex filament is over 5 miles in length.

As shown in Table 3.4, fiber production in the United States is on the order of 9 billion pounds annually. The value of the U.S. fiber industry is in the range of tens of billions of dollars, an amount that is multiplied manyfold by the time the fibers reach the consumer market in products ranging from ropes to textiles to automobile tires. Some typical applications of fiber are listed in Table 3.5.

Over the past several decades, a number of trends have become evident in the commodity fiber business:

TABLE 3.4 U.S. Production of Synthetic Fibers, 1992

	Pounds (billions)
Acetate	0.50
Acrylic	0.44
Nylon	2.55
Olefin	1.99
Polyester	3.58

SOURCE: Data from *Chemical & Engineering News* (1993).

TABLE 3.5 Typical Applications of Fiber

Use	Polymer
Clothing	Polyester, nylon, polypropylene, acrylics, spandex
Carpets	Nylon, polypropylene, polyester
Protective garments	Aramids, fluoropolymers, polyethylene
Synthetic paper	Polyethylene
Ropes, cables	Polypropylene, nylon, aramids
Golf shafts, fishing rods	Carbon, aramids
Cement reinforcement	Polyethylene
Brake linings	Aramids, acrylics

- Synthetic fiber volumes have grown at the expense of natural fibers. The drivers are lower costs and technical improvements, which allow the synthetics to emulate desirable natural fiber aesthetics while exhibiting superior in-use performance.
- The commodity markets are divided primarily among nylon, polyester, and polyolefin, with polyester emerging as the largest. Cost-performance and environmental considerations have led to a diminution in the use of cellulosics and acrylics.
- The introduction of a new commodity fiber is generally regarded as unlikely.

This same time period has seen the rapid growth of high-performance fiber technologies. These technologies fall into three classes:

- High-modulus, high-strength fibers based on rodlike, liquid crystalline (nematogenic) polymers. The most common examples are the lyotropic aramids and the thermotropic polyesters. These fibers are characterized by tensile moduli greater than 70 gigapascals (GPa), tensile strengths on the order of 3 to 4 GPa, and low properties in compression or shear.
- Morphological manipulation of conventional polymers, such as high-mo-

lecular-weight polyethylene or poly(vinyl alcohol), to achieve fibers with levels of mechanical performance similar to those of the liquid crystalline polymer fibers.

- Polymeric precursor fibers that can be converted to other chemical forms after spinning. The most common examples are acrylic fibers that can be converted to carbon fibers and a variety of silicon-containing polymeric fibers that can be converted to silicon carbide or silicon nitride fibers.

Typical applications of high-performance fibers are composite reinforcement, ropes and cables, and antiballistic clothing. As a group, these fibers represent successful technical developments, but they have proved less commercially attractive than once believed for a variety of reasons.

The spinning process can be described as follows. A polymer is first converted to a liquid through melting or dissolution, and the liquid is then continuously forced through a spinnerette (a plate with many of small holes) to form filaments. Most polymeric fibers are semicrystalline. If the polymer forms a stable melt, the process is called melt spinning. For polymers that degrade prior to melting, the polymer is spun from a solution; if the solvent is evaporated, the process is termed dry spinning; if the solution is coagulated in a nonsolvent bath, the process is termed wet spinning. Removal of the spinnerette from the wet spinning coagulation bath is the innovation known as dry-jet wet spinning. In practice, the spinnerette holes are usually much larger than the desired final filament diameter; hence the filaments are stretched either during and/or after the spinning process. The ratio of final filament velocity to the initial filament velocity is termed the drawdown ratio. The principal parameters controlling the as-spun structure and, hence, properties of the as-spun filament are the rate of cooling and the applied stress. These parameters control the extent of molecular chain orientation (time to orient/time to relax) and the degree of crystallinity achieved during spinning. Crystallinity once formed can be further oriented by stretching and perfected through annealing. Key structural elements are the amount and orientation of crystalline regions, the orientation of noncrystalline regions, and connectivity between regions, tie molecules, and so on.

Careful control of the sequence in which chains are oriented and crystallized has a profound effect on the microstructure produced. Such controlled processing allows, for example, the decoupling of crystalline and noncrystalline orientation, enabling fibers with high tensile modulus (correlated with high crystalline orientation) and low thermal shrinkage (correlated with low noncrystalline orientation) to be produced. Typical spinning speeds are thousands of meters per minute, typical melt drawdowns are on the order of 100, and typical solid-state draw ratios range from about 2 to 6 in conventional processing to greater than 50 in the production of certain high-performance products. High-performance fiber processing is characterized by maximizing axial chain orientation and minimizing structural defects formed during spinning, giving rise to mechanical performance that approaches theoretical limits. To control friction and static behavior in subsequent processing, a variety of oils or other surface treatments are applied to the fibers prior to take-up. The many complex processing steps of fibers add to the stress-temperature history of the fiber and hence significantly modify the end-use properties of the material.

To a large extent, the conditions employed in spinning, in addition to the particular chemistry of the polymer being spun, determine the end-use performance of a fiber. Work on future fibers will focus on producing cost-performance improvements and product variants through morphological control rather than new chemistries. With the huge lengths of fibers produced, process robustness and property uniformity have always been major issues; future products will make more use of advanced computerized process control and will operate in areas of property response that are less sensitive to minor process variation. Elimination of downstream process steps will lead to additional cost-performance improvements, for example, on-line texturing and surface modifications to meet specific friction or adhesion requirements.

Environmental considerations will influence future fiber developments in a number of areas. The elimination of solvent-based processing will be driven by stricter emissions standards, as will the elimination of heavy metal catalysis. Novel processes based on very fast melting techniques (e.g., RF heating or lasers) or the use of supercritical carbon dioxide as a polymerization and/or spinning solvent will become more commercially attractive. The reduction of off-specification production will become more important as the cost of waste disposal increases and as easy-to-reclaim fibers grow in importance (e.g., biodegradable cellulosics produced without organic solvents, or poly(ethylene terephthalate) reduced to usable monomers in a process that has become a commercial reality).

The future of high-performance fibers lies in the reduction of costs and the improvement of utilization. The former is best influenced by lower-cost monomers, and the latter through the development of manufacturing technologies that allow cost-effective part production from fiber-reinforced composites. High-performance fiber development will cease to be solely performance driven and will, as in the case of all other fibers, become driven by cost and performance. Silks, produced by worms and spiders, have attracted attention because they possess tensile properties similar to those of high-performance synthetic fibers but with much higher toughness. The use of recombinant DNA techniques allows silks of specific molecular architectures to be produced and their performance to be correlated with specific chemical and physical features. The increased structure-property insights gained from these studies should allow the definition of biomimetic fibers, based on other than naturally occurring amino acids, with greatly improved performance characteristics.

Adhesives

An adhesive is a material that, by means of surface attachment, can hold together solid materials. Adhesives have been used for most of recorded history. They are mentioned in Egyptian hieroglyphics, in the Bible, and in the writings of the early natural philosophers. The physical strength of an assembly made by the use of adhesives, known as an adhesive joint, is due partly to the forces of adhesion, but primarily to the cohesive strength of the polymeric materials used to formulate the adhesive. Thus, the range of strengths available in adhesive joints is limited to the strengths of the polymers useful in the formulation of adhesives. Indeed, the technology of adhesives tracks well with the technology of polymers. As new polymers were synthesized, new adhesives were developed that used those polymers.

Adhesives are typically classified by their use or application. Thus structural adhesives are those materials used to join engineering materials such as metals, wood, and composites. Usually, it is expected that an adhesive joint made with a structural adhesive is capable of sustaining a stress load of 1,000 psi (6.9 MPa) for extended periods of time. Hot melt adhesives are those adhesives that are applied from the melt and whose properties are attained when the adhesive solidifies. Pressure-sensitive adhesives provide adherence and strength with only finger pressure during application. Adhesive tapes are manufactured by applying a pressure-sensitive adhesive to a backing. Rubber-based adhesives are, as the name implies, based on elastomers and are usually applied as a mastic or spray applied from solvent or water. Pressure-sensitive adhesives can be considered to be a subset of rubber-based adhesives.

The ease of application of pressure-sensitive adhesives is superior to all other types of adhesives except possibly hot melt adhesives. Responsivity to finger pressure alone forming a bond is a desirable property, and pressure-sensitive adhesives of sufficient strength to perform structural tasks have been developed recently. One of the major uses of these double-coated foam tapes is to fasten most of the exterior and interior decorative and semistructural materials to the body of an automobile. The use of these foam tapes allows faster assembly and eliminates mechanical fasteners, which are a source of corrosion.

Each of the major classes of adhesives described above can be further classified by its chemistry. Thus, the majority of structural adhesives are based on one or more of the following chemistries: phenolic, epoxy, acrylic, bismaleimide, imide, and protein (derived from blood, soybean, casein, and so on). The majority of hot melt adhesives are based on one or more of the following chemistries: waxy hydrocarbons, polyethylene, polypropylene, ethylene-vinyl acetate, polyamides, and polyesters. Rubber-based adhesives are, for the most part, formulated using neoprene, nitrile, and natural rubbers. Pressure-sensitive adhesives are based on natural rubber, vinyl ethers, acrylics, silicones, and isoprene-styrene block co-polymers. Many paper-binding adhesives are based on dextrin or other starch-based materials. "White glue," used for wood bonding, is a poly(vinyl acetate) emulsion.

Adhesives have several advantages over other joining technologies. In general, adhesives have a lower density than mechanical fasteners, and so weight savings can be realized. Polymer-based adhesives have viscoelastic character and are thus capable of energy absorption. The energy absorption manifests itself in the form of dampening of vibrations and in the increase of fatigue resistance of a joint. Adhesives can be used to join electrochemically dissimilar materials and provide a corrosion-resistant joint. Adhesive joining is limited by the fact that an engineering database is unavailable for most adhesive materials.

The strength and durability of an adhesive bond are subject to the nature of the surfaces to be joined. Part of the reason industrial adhesives have been so successful is that methods have been found to clean and treat surfaces to form good bonds. A better understanding of proper surface preparation for adhesives is needed. The major limitations to the broader use of adhesives in industry are the extreme sensitivity of adhesive bonding to surface conditions and the lack of a nondestructive quality control method.

Adhesive technology can be solidly advanced by the synthesis of new monomers and polymers that extend the range of applicability of adhesive bonding. Thus, new materials should allow adhesives to be more flexible at cryogenic temperatures, more oxidation resistant at high temperatures, stronger at elevated temperatures, and more tolerant of an ill-prepared or low-surface-energy adherent. The engineering aspects of adhesive technology can be solidly advanced by including adhesive technology in university engineering courses and establishing an engineering database. In addition, an easy, nondestructive method of predicting the strength of a joint would be a major advance in the applicability of adhesives. Two drivers of advances in adhesive technology in the near future are economics and the environment. To be environmentally acceptable, new adhesive formulations should contain a minimum of solvent and in some applications should be biodegradable. To be economically attractive, adhesives should be easy to use and should provide a value-added feature to the customer that outweighs the disadvantages cited above.

Blends and Alloys

The time scale for introducing totally new polymers is increasing because the simplest monomers and the processes for converting them into polymers have already been identified and introduced into the marketplace. Furthermore, with increasing regulatory obstacles and the high cost of research, the economic stakes for introducing generically new polymers based on previously unknown chemistry and manufacturing processes have been raised considerably. Because this field was initially dominated by the ready opportunities for chemical innovation, serious development based on the more physical approach of alloying or blending existing polymers did not begin until the late 1970s and 1980s. Now, the area of polymer blends is one of the routes to new materials that is most actively pursued by the polymer industry.

There are several driving forces for blending two or more existing polymers. Quite often, the goal is to achieve a material having a combination of the properties unique to each of the components, such as chemical resistance and toughness. Another issue is cost reduction; a high-performance material can be blended with a lower-cost polymer to expand market opportunities. A third driving force for blending polymers of different types is addition of elastomeric materials to rigid and brittle polymers for the purpose of toughening. Such blends were the first commercial example of polymer blend technology and, even today, probably account for the largest volume of manufacturing of multicomponent polymer systems. The main problem is that frequently when polymers are blended, many critical properties are severely depressed because of incompatibility. On the other hand, some blends yield more or less additive property responses, and others display certain levels of synergism. The problem is knowing how to predict in advance which will occur and how to remedy deficiencies.

From a fundamental point of view, one of the most interesting questions to ask about a blend of two polymers is whether they form a miscible mixture or solution. The thermodynamics of polymer blends is quite different from that of mixtures of low-molecular-weight materials, owing to their molecular size and the greater importance of compressibility effects. Because of these, miscibility of two polymers generally is driven by energetic rather than the usual entropy considerations that cause most low-molecular-weight materials to be soluble in one another. The simple theories predict that miscibility of blends is unlikely; however, recent research has shown that by carefully selecting or designing the component polymers there are many exceptions to this forecast. The phase diagram for polymer blends is often opposite of what is found for solutions of low-molecular-weight compounds. Polymers often phase separate on heating rather than on cooling as expected for compounds of low molecular weight. Theories to explain the behavior of miscible polymer blends have emerged, but theoretical guidance for predicting the responsible interactions is primitive. With the advent of modern computing power and software development, molecular mechanics calculations of this type are being attempted. Neutron scattering has provided considerable insight about the thermodynamic behavior of blends and the processes of phase separation.

One of the earliest blend products was a miscible mixture of poly(phenylene oxide) and polystyrene. The former is relatively expensive and rather difficult to process. The addition of polystyrene lowers the cost and makes processing easier. Numerous other commercial products are now based on miscible or partially miscible polymer pairs, including polycarbonate-polyester blends and high-performance ABS materials.

For mixtures that are not miscible, the most fundamental issues relate to phase morphology and the nature of the interface between these phases. Frequently, the unfavorable polymer-polymer interactions that lead to immiscibility cause an unstable and uncontrolled morphology and a weak interface. These features translate into poor mechanical properties and low-value products, that is, incompatibility. When this is the case, strategies for achieving compatibility are sought, generally employing block or graft copolymers to be located at the interface, much like surfactants. These copolymers can be formed separately and added to the blend or formed in situ by reactive coupling at the interface during processing. The former route has, for example, made it possible to make blends of polyethylene and polystyrene useful for certain packaging applications by addition of block copolymers formed via anionic synthesis. However, viable synthetic routes to block copolymers needed for most commercially interesting combinations of polymer pairs are not available. For this reason, the route of reactive compatibilization is especially attractive and is receiving a great deal of attention for development of commercial products. It involves forming block or graft copolymers in situ during melt processing by reaction of functional groups. Extensive opportunities exist for developing schemes for compatibilization and for fundamental understanding of their mechanisms. A better understanding of polymer-polymer interactions and interfaces (e.g., interfacial tension, adhesion, and reactions at interfaces) is essential. Especially important is the development of experimental techniques and better theories for exploring the physics of block and graft copolymers at such interfaces. This knowledge must be integrated with a better understanding of the rheology and processing of multiphase polymeric materials so that the morphology and interfacial behavior of these materials can be controlled.

A wide variety of compatibilized polymer alloys have been commercialized, and the area is experiencing a high rate of growth. A product based on poly(phenylene oxide), a polyamide, and an elastomer has been introduced for use in forming injection-molded automobile fenders and is currently being placed on several models of U.S. and European-made automobiles. The polyamide confers toughness and chemical resistance, the poly(phenylene oxide) contributes resistance to the harsh thermal environment of automotive paint ovens, while the elastomer provides toughening. Another automotive application is the formation of plastic bumpers by injection molding of ternary blends of polycarbonate, poly(butylene terephthalate), and a core shell emulsion-made elastomeric impact modifier (Figure 3.5). In this blend, the polycarbonate brings toughness, which is augmented at low temperatures by the impact modifier, while the poly(butylene terephthalate) brings the needed chemical resistance to survive contact with gasoline, oils, and greases. In the first example, the poly(phenylene oxide) and polyamide are very incompatible, and reactive coupling of the phases is required for morphology control and for interfacial strengthening. In the second example, the polycarbonate and polyester apparently interact well enough that no compatibilizer is needed.

FIGURE 3.5 High-performance alloys are now used extensively in the demanding exterior body applications of automobiles. For example, front fenders are molded from a PPO-nylon alloy that can withstand paint oven temperatures of 400°F and above. They are chosen for their class-A surface, dimensional stability, impact strength, and corrosion and chemical resistance. The side claddings on these vehicles are molded of a resin that is a polyester-polycarbonate alloy, chosen for its cold temperature impact strength, chemical resistance, and quality surface. For the front and rear bumper fascias, a copolyester elastomer is used, characterized by its cold temperature ductility, class-A surface, chemical resistance, and on-line paintability at temperatures exceeding 280°F. More than 60 pounds of engineering thermoplastics can be found on many of the vehicles. SOURCE: Photograph courtesy of GE Plastics, Southfield, Michigan.

Toughening by the addition of rubber was first practiced for commodity polymers, such as polystyrene, poly(vinyl chloride), polypropylene, and poly(methyl methacrylate) (PMMA). Widely different processes and product designs were required to achieve optimal products. Now this approach is being applied to engineering thermoplastics and thermosets in order to move these materials into applications that require stringent mechanical performance under demanding conditions. This ensures an excellent growth opportunity for a variety of toughening agents. Elastomers with low glass transition temperatures are needed to impart toughness at low use temperatures, while thermal and oxidative resistance are needed to survive the high temperatures required for processing these materials. In addition, these elastomers must be dispersed within the matrix to an appropriate morphology (or size scale) and adequately coupled to the matrix. These two issues are often interrelated and specific to the particular matrix material. Continued efforts will be required to produce a better understanding of the various toughening mechanisms that are applicable to engineering polymers.

Numerous opportunities exist to achieve better understanding that would shorten the time to develop new blends and alloys. There is an interesting parallel between this field and alloying in metallurgy, and the polymer community may be able to learn from the long experience of metallurgists. Both fields involve a broad spectrum of issues including synthesis, processing, physical structure, interfaces, fracture mechanics, and lifetime prediction. The United States is currently in a position of technical leadership; however, companies and universities around the world are also aggressively pursuing research and development in this field.

Structural Composites

Polymer composites can provide the greatest strength-to-weight and stiffness-to-weight ratios available in any material, even the lightest, strongest metals. Hence, high-performance and fuel-economy-driven applications are prime uses of such composites. One of the most important attributes is the opportunity to design various critical properties to suit the intended application. Indeed, performance may be controlled by altering the constituents, their geometries and arrangement, and the interfaces between them in the composite systems. This makes it possible to "create" materials tailored to applications, the single greatest advantage and future promise of these material systems. Structural composites are of interest in aerospace applications and in numerous industrial and consumer uses in which light weight, high strength, long fatigue life, and enhanced corrosion resistance are critical. Much needs to be done to advance processibility and durability, to provide a more comprehensive database, and to improve the economics of these systems. A wide range of future needs encompasses synthesis, characterization, processing, testing, and modeling of important polymer matrix composite systems.

In general, the future of polymer matrix composites is bright. The engineering community is now in the second generation of applications of composites, and primary structures are now being designed with these materials. There is a growing confidence in the reliability and durability of polymer composites and a growing realization that they hold the promise of economic as well as engineering gain. Commercial programs such as high-speed civil transport will not succeed without the use of polymer composites. Integrated synthesis, processing, characterization, and modeling will allow the use of molecular concepts for the design of the material system and to estimate the effect of how the materials are put together on the performance, economy, and reliability of the resulting component. A more precise understanding of the manufacturing, processing, and component design steps will greatly accelerate the acceptance of these advanced materials. New horizons for properties and performance, for example, in smart and intelligent materials, actuators, sensors, high-temperature organic materials, and multicomponent hybrid systems, will involve the potential of introducing a new age of economic success and technical excellence. It has been estimated that finished-product businesses of greater than $5B annually already exist for the aggregate of polymers, reinforcements, prepregs, tooling machinery, and other ancillary products (McDermott, 1993).

Advanced polymer matrix composites have been used for more than 20 years, for example, on the B-1 bomber and for many top-of-the-line Navy and Air Force jet fighters. For military purposes, the high performance and stealthiness of composites have often outweighed issues of durability and even safety. Building lighter, more maneuverable tanks, trucks, and armored vehicles might be an area for future military growth. However, as the Pentagon's budget shrinks, efforts to transform these materials into civilian uses are under way (Pasztor, 1992). Problems include the need to identify significant nondefense companies that will use advanced composites. For nearly 30 years, it has been suggested that aircraft designers around the world would rapidly utilize these new materials. Unfortunately, those predictions have not been realized, and U.S. plants making polymer matrix composites are now operating at less than 50 percent of capacity. For a number of reasons, there is continued reticence to employ these advanced materials in many areas, particularly in commercial aviation. Costs, processibility, and durability appear to be the major issues. To this point, this area has been considered a technical success but not a financial success. Nevertheless, aircraft in various stages of development have composites as some fraction of their structural weight. For example, 15 percent of the Boeing 777, 6 percent of the MD-11 Trijet, and 15 percent of the MD-12 are estimated to be composites. European aviation firms have begun flight-testing an all-composite tail rotor for a helicopter, and Japanese efforts are under way to develop a military helicopter that has a very high composite content.

It has been predicted that in the future, fiber-reinforced composites (FRCs) will partially replace conventional materials in civil engineering applications. These could include buildings, bridges, sewage and water treatment facilities, marine structures, parking garages, and many other examples of infrastructure components. Composite materials are also expected to help replace conventional materials such as steel and concrete in many future projects. A volume of $3T for fiber-reinforced composites in the rehabilitation of the country's infrastructure has been estimated (Barbero and Gangarao, 1991). The polymer matrix resin composites discussed above have already made inroads in areas such as antenna coverage and water treatment plants. Less expensive fiber-reinforced plastics (FRPs), such as unsaturated polyester styrene matrix systems reinforced with glass fibers, have become important automobile and bathroom construction materials. Sheet molding compounds, which are used extensively in automobiles and housing, are not considered by many structural engineers to be suitable for infrastructure replacement owing to their relatively low strength. Advanced polymer composites, on the other hand, which often consist of continuously reinforced fiber materials, have superior strength and stiffness.

Liquid Crystalline Polymers

The liquid crystalline nature of stiff polymer molecules in solution was predicted by Onsager in 1947, further refined by Flory in 1956, and experimentally verified through aramid investigations at the Du Pont Company in the 1960s. Flory suggested that as the molecular chain becomes more rodlike, a critical aspect ratio is reached, above which the molecules necessarily line up to pack efficiently in three dimensions. Liquid crystal polymer concepts have been extended to encompass a vast number of homopolymer and copolymer compositions that exhibit either lyotropic or thermotropic behavior. Industrially, most of the effort has been focused on the main-chain nematic polymers. These polymers combine inherently high thermal and mechanical properties with processing ease and versatility. Processing ease originates from the facile way that molecular rods can slide by one another, the very high mechanical properties come from the "extended chain" morphology present in the solid state, and the thermal stability derives from the highly aromatic chain chemistry. Inherent in this structure is a high level of structural, and hence property, anisotropy (for example, the axial modulus is 1 to 2 orders of magnitude higher than the transverse modulus). The direction of molecular chain orientation is coincident with the direction of covalent bonding in the chain; normal to the orientation direction the bonding is secondary (van der Waals, hydrogen bonding, and so on). Low orientation in these materials means global but not local randomness, and properties within "domains" are highly anisotropic.

A useful spin-off of the study of liquid crystal polymers was the recognition of the importance of mesophases in the development of structure in conventional polymers. Examples of this include the stiffening of polyimide backbones to reduce the expansion coefficient and improve processibility and the recognition of the importance of a pseudo-hexagonal (rotator, transient nematic) phase in the crystallization of oriented polymer melts. Increasing the end-to-end distance of conventional polymers through the application of either mechanical or electromagnetic fields can lead to the formation of structure equivalent to that achieved by the manipulation of molecularly stiff molecules.

Fibers from lyotropic para-aramid polymers (Figure 3.6) were initially commercialized by the Du Pont Company in 1970 under the Kevlar® trademark. The fibers are dry-jet wet spun from 100 percent sulfuric acid solution with sufficient

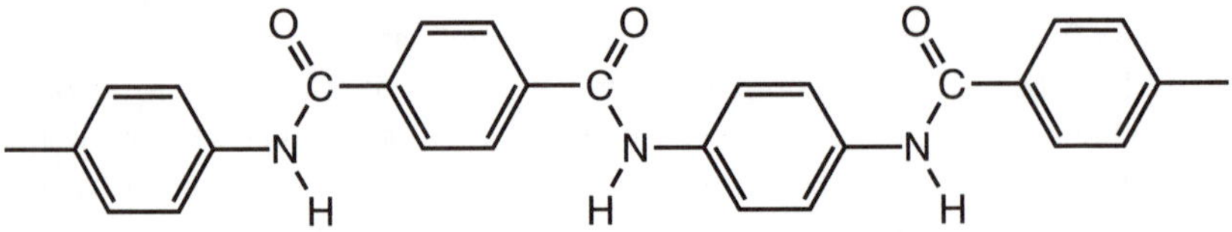

FIGURE 3.6 A lyotropic para-aramid liquid crystal polymer.

drawdown to orient the molecular chains, actually the liquid crystal domains, parallel to the fiber axis. An annealing step may be performed to improve structural perfection, resulting in an increase of fiber modulus. These fibers have very high modulus and tensile strengths as well as excellent thermal and environmental stability. Weaknesses include low compressive properties (endemic with all highly uniaxially oriented polymers) and a significant moisture regain. Worldwide fiber production capacity is about 70 million pounds (1991). Selling prices vary according to grade (i.e., modulus level) and market, ranging from as low as $8 per pound to over $50 per pound. Consumption worldwide in 1990 was about 50 million pounds, somewhat trailing capacity. Major markets include reinforcement for rubber and composites, protective apparel, ropes and cable, and asbestos replacement. The use of para-aramid fiber is projected to grow at greater than 10 percent per year worldwide over the next 5 years. The environmental issues involved in the handling and disposal of large quantities of sulfuric acid or other solvents may make thermotropic approaches more attractive in the future.

During the 1980s, thermotropic copolyesters were commercialized worldwide. More versatile than the lyotropic polymers, these nematic copolyesters (Figure 3.7) are amenable to uniaxial processing, such as fiber formation, and three-dimensional processing, such as injection molding, utilizing essentially conventional thermoplastic processing techniques. While fiber products exist, most of the commercial thermotropic copolyester is sold as glass- or mineral-filled molding resins, the majority into electrical and electronic markets. U.S. volume in 1991 was about 4.3 million pounds, at an average selling price of about $8 per pound. As in the case of the aramids, thermal and environmental stability is excellent. Advantages of these molding resins are the extremely low viscosity, allowing the filling of complex, thin-walled molds, excellent mold reproduction because of the low change in volume between liquid and solid, and fast cycle times. Weaknesses include property anisotropy and high cost.

The future growth of the main-chain nematogenic polymers will be dominated by two factors:

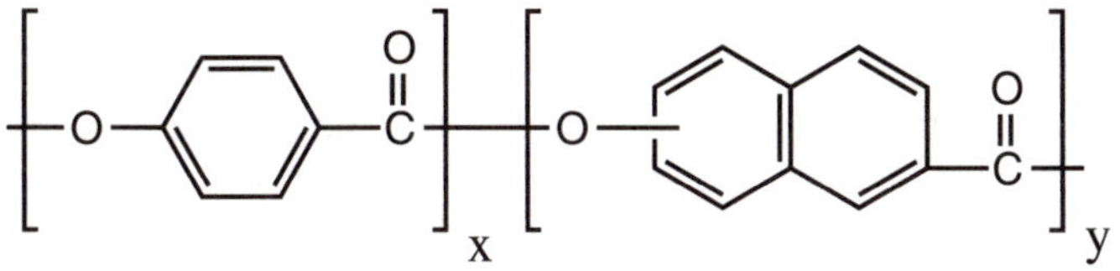

FIGURE 3.7 An aromatic polyester thermotropic liquid crystal polymer.

- Polymer cost (ultimately dependent on monomer cost), and
- Processing technology allowing cost-effective exploitation of properties, including orientation control in finished parts, as well as new forms (e.g., films, shaped extrudates, nonwovens, foams, and multilayers).

Two particularly intriguing properties of nematogenic polymers not yet important commercially are ductility under cryogenic conditions and very low permeabilities of small molecules through the solid-state structure (high barrier properties).

A potentially attractive route to both lower price and improved property control is the blending of liquid crystal polymers with conventional polymers. An extensive literature exists, and interesting concepts such as self-reinforcing composites and molecular composites have been developed to describe immiscible and miscible liquid crystal polymer-containing blends. Major problems encountered in this technology include:

- Inherent immiscibility of mesogenic and conventional polymers, leading to large-scale phase separation;
- Strong dependence of blend morphology (properties) on processing and polymer variables; and
- Lack of adhesion between phases.

To date, commercial success for such blends has proved elusive. A related approach is the use of liquid crystal polymers in conventional composites, either as reinforcing fiber, matrix, or both. Penetration into conventional composite markets has been slow, the major problems being poor adhesion, poor compression (fibers), and the lack of design criteria for composite parts where both matrix and ply are anisotropic.

The potential of polymeric liquid crystals in device rather than structural applications has been recognized in both industry and academia, but no commercially viable products have yet emerged. The combination of inherent order, environmental stability, and ease of processing has led to interest in the use of polymeric liquid crystalline textures in applications as diverse as nonlinear optics, optical data storage, and "orienting carriers" for conducting polymers. With structural parameters of secondary importance, all textures are under active investigation. Both main-chain and side-chain approaches are of interest, the goal being to tailor orientational and transitional states to the specific end use desired. Emerging problems include achieving sufficient density of active species to produce materials with competitive figures of merit (i.e., dipole concentration for nonlinear optical applications) and balancing mesogenicity effects with high and stable use temperatures. Clearly, the introduction of mesogenicity into polymers opens vast possibilities for molecular design, which may ultimately lead to the creation of materials with highly specific and unique property sets.

Films, Membranes, and Coatings

Polymers are used in many applications in which their main function is to regulate the migration of small molecules or ions from one region to another. Examples include containers whose walls must keep oxygen outside or carbon dioxide and water inside; coatings that protect substrates from water, oxygen, and salts; packaging films to protect foodstuffs from contamination, oxidation, or dehydration; so-called "smart packages," which allow vegetables to respire by balancing both oxygen and carbon dioxide transmission so that they remain fresh for long storage or shipping times; thin films for controlled delivery of drugs, fertilizers, herbicides, and so on; and ultrathin membranes for separation of fluid mixtures. These diverse functions can be achieved partly because the permeability to small molecules via a solution-diffusion mechanism can be varied over enormous ranges by manipulation of the molecular and physical structure of the polymer. The polymer that has the lowest known permeability to gases is bone-dry poly(vinyl alcohol), while the recently discovered poly(trimethylsilyl propyne) is the most permeable polymer known to date. The span between these limits for oxygen gas is a factor of 10^{10}. A variety of factors, including free volume, intermolecular forces, chain stiffness, and mobility, act together to cause this enormous range of transport behavior. Recent experimental work has provided a great deal of insight, while attempts to simulate the diffusional process using molecular mechanics are at a very primitive stage. There is clearly a need for guidance in molecular design of polymers for each of the types of applications described in more detail below. In addition, innovations in processing are needed.

Barrier Polymers

As shown earlier, packaging applications currently consume roughly one-third of the production of thermoplastic polymers for fabrication of a wide array of rigid and flexible package designs (see Figure 3.2). These packages must have a variety of attributes, but one of the most important is to keep contaminants, especially oxygen, out, while critical contents such as carbon dioxide, flavors, and moisture are kept inside. Metals and glass are usually almost perfect barriers, whereas polymers always have a finite permeability, which can limit the shelf-life of the products they protect. In spite of this deficiency, the light weight, low cost, ease of fabrication, toughness, and clarity of polymers have driven producers to convert from metal and glass to polymeric packaging. Polymers often provide considerable savings in raw materials, fabrication, and transportation, as well as improved safety for the consumer relative to glass; however, these advantages must be weighed against complex life-cycle issues now being addressed. The following discussion illustrates the current state of this technology, its problems, and future opportunities.

There are certain polymer molecular structures that provide good barrier properties; however, these structural features seem invariably to lead to other problems. For example, the polar structures of poly(vinyl alcohol), polyacrylonitrile, and poly(vinylidene chloride) make these materials extremely good barriers to oxygen or carbon dioxide under certain conditions, but each material is very difficult to melt fabricate for the same reason. The good barrier properties stem from the strong interchain forces caused by polarity that make diffusional jumps of penetrant molecules very difficult. To overcome these same forces by heating, so that the polymer chains can move in relation to one another in a melt, requires temperatures that cause these reactive materials to degrade chemically by various mechanisms. Thus neither poly(vinyl alcohol) nor polyacrylonitrile can be melt processed in its pure form. Resorting to solvent processing of these materials or using them to make copolymers compromises their value. Poly(vinyl alcohol), by virtue of its hydrogen bonding capability, is very  hygroscopic, to the point of being water soluble, and this property prevents its use as a barrier material in the pure form even if it could be melt processed. In general, polarity favors good oxygen barrier properties but leads to poor water barrier properties. This is true for aliphatic polyamides (nylon). On the other hand, very nonpolar materials, such as polyethylene and polypropylene, are excellent barriers to water but not oxygen. This property-processibility trade-off has led to an interest in composite structures. The "composites" can be at the molecular level (copolymers), microlevel (blends), or macrolevel (multilayers).

The attractive barrier characteristics of poly(vinyl alcohol) have been captured via copolymers, and this achievement has led to some important commercial products using clever molecular engineering and processes that minimize its shortcomings of water uptake and lack of melt processibility. Copolymers containing units of ethylene and vinyl alcohol are made commercially by starting with ethylene and vinyl acetate copolymers and then hydrolyzing them. By critically balancing the structure of these materials, melt processible products that are relatively good barriers with reduced moisture sensitivity can be achieved. These copolymers are incorporated into multilayer structures by coextrusion processes. For example, blow-molded bottles with five to seven layers in the side wall are in commercial use for marketing very sensitive foodstuffs. Lightweight, squeezable, fracture-resistant bottles for ketchup are now on the market. Layers of ethylene/vinyl alcohol copolymer provide the oxygen barrier to prevent spoilage of the sensitive tomato base, while layers of polypropylene provide the water barrier that prevents dehydration on storage. Interlayers are often needed to adhere the functional layers to one another when the two differ greatly in chemical structure. Sometimes a mixed layer is included to accommodate recycled material from the process. The barrier function can also be provided by metal foil or by coatings of other polymers or inorganic layers onto containers. Of course, composite structures are inherently more difficult to recycle. Layers based on halogen-based polymers generate acid gases upon incineration. Reconciling these issues will be a major preoccupation during the next decade.

One of the major developments over the past two decades has been the replacement of glass with plastics in bottles for soft drink merchandising. The driving forces for this conversion were issues of cost, weight, safety, and total energy considerations. The commercialization of this technology using poly(ethylene terephthalate), or PET, involved innovative developments in processing for increasing molecular weight (solid-state reaction) and for fabrication (injection-blow molding) to achieve a highly oriented and transparent bottle. The carbon dioxide permeability of PET provides just enough shelf-life for very successful marketing of large 2-liter products; however, smaller bottles, such as the half liter, with a higher surface-to-volume ratio, have a shorter shelf-life. PET is also easily recycled, and considerable progress is being made in this area. PET, however, has not been able to succeed so far in the beer packaging market, owing to marginal oxygen barrier characteristics among other issues. Polyesters with much better properties are known, such as poly(ethylene naphthalene-2,6-dicarboxylate), but these have not yet become commercial because economical processes for raw material production have not been developed.

Current areas of focus include the development of packages that can be directly microwaved, such as packages for soups in single-serving sizes, and controlled atmosphere packaging, which is capable of keeping fruits and vegetables fresh for weeks. Successes in the latter area could revolutionize the agriculture and food industries of the world in terms of where produce is grown, how it is distributed, and who has access to it. There are some clear fundamental challenges for development of new barrier materials that are economical, melt processible, and environmentally friendly, but significantly better than current ones in terms of permeability to oxygen, water, and oil.

Membranes

Membrane-based processes that provide many useful functions for society, usually at lower cost, particularly in terms of energy, have achieved substantial commercial importance relatively recently. The majority of the membranes used are made from polymers. It has been estimated that the worldwide sales of membranes in 1990 was of the order of $2B (ancillary hardware associated with membrane systems raises this value severalfold) and that this industry is growing at 12 to 15 percent per year (Strathman, 1991). The United States is clearly in the lead position, but Europe and Japan are gaining rapidly. There is interest in other materials, such as ceramics, but it is clear that polymers will dominate in most uses. For the most part, the major limitation of membrane technology is the performance of the membrane itself; hence, sustained growth demands new developments in membrane materials and membrane fabrication.

Membranes are used to produce potable water from the sea and brackish waters, to treat industrial effluents, to recover hydrogen from off-gases, to produce nitrogen and oxygen-enriched air from air, to upgrade fuel gases, and to purify molecular solutions in the chemical and pharmaceutical industries. They are the key elements in artificial kidneys and controlled drug delivery systems. Basically, membranes may function in one of two general ways, depending on the separation to be performed and the structure of the membrane.

Some membranes act as passive filters, albeit usually on a very small scale. These membranes have pores through which fluid flows, but the pores retain larger particles, colloids, or macromolecules (e.g., proteins). Depending on the scale of the pores and the solute or particles, the operations are subdivided into ultrafiltration, microfiltration, and macrofiltration. The material dictates the manner in which the membrane can be formed and especially the size and distribution of the pores. Porous polymer-based membranes are made by solution processes, mechanical stretching, extraction, or ion bombardment processes. The nature of the membrane material is a key factor in resistance to damage and fouling and in compatibility with the fluid phase (e.g., blood).

When the membrane is nonporous, the polymer is a more direct participant in the transport process. Permeation across the membrane involves dissolution of the penetrant into the polymer and then its diffusion to the other surface, that is, a solution-diffusion mechanism. The thermodynamic solubility and kinetic diffusion coefficients of penetrants in polymers depend critically on the molecular structures of the penetrant and the polymer and their interactions. This is the mechanism by which reverse osmosis, gas separation, and pervaporation membranes function.

In order to have usefully high rates of production in membrane processes, it is generally necessary to have membranes that are very thin and to have a very large membrane area packaged in small volumes. Ingenious approaches have been developed to achieve both. Membranes may be in the form of a flat sheet wrapped into a spiral for packaging into modules or in the form of very fine hollow fibers. In either case the membrane has a dense skin that is very thin (0.01 to 2 micrometers [μm]), which is the selective rate-controlling element that overlays a much thicker porous substructure to provide mechanical support or integrity at minimal permeation resistance. The skin and the substructure may be integral, made of the same material. The method to fabricate such asymmetric membranes was discovered in the 1960s and was first applied to make reverse osmosis membranes and later to make gas separation membranes. This was the pivotal invention that has made most of modern membrane technology possible. A variety of composite membrane concepts were developed later that have the advantage that the skin and porous support are not integral and can, in fact, be made of different materials. This is especially useful when the active skin material is very expensive. Reverse osmosis and gas separation membranes of both types are in current use.

There is growth in almost all sectors of the membrane industry; however, the opportunities for future impact by new polymer technology appear somewhat uneven. For example, one of the major limitations to the use of ultrafiltration-type processes in the growing biotechnology arena is the tendency for surface fouling by protein and related macromolecules. The discovery of new membrane materials or surface treatments that solve this problem would be of major importance.

Intense polymer research related to reverse osmosis during the 1960s and 1970s led to commercial installation of desalinization plants around the world. Membranes in use are made of cellulose acetate and polyamides. Future demands for fresh water from the sea could stimulate renewed research interest in this area. Currently most of the efforts are devoted to developing reverse osmosis membranes and processes for removal of organic pollutants, rather than salt, from water.

Gas separation is clearly one of the most active and promising areas of membrane technology for polymer science and engineering (Figure 3.8). The first commercial membranes introduced in the late 1970s were hollow fibers formed from polysulfone by using a unique technology to remove minute surface defects. Since then, other polymers have been introduced in the United States, including cellulose acetate, polydimethylsiloxane (PDMS), ethyl cellulose, brominated polycarbonate, and polyimides. The first materials selected for this purpose were simply available commercial polymers that had adequate properties. New generations of materials especially tailored for gas separation are being sought to open new business opportunities. The key issues involve certain trade-offs. The polymer must be soluble enough to be fabricated into a membrane, but it needs resistance to chemicals that may be in the feed streams to be separated. The membrane should have a high intrinsic permeability to gases in order to achieve high productivity, but the permeation should be selective; that is, one gas, for example, O_2, must permeate much faster than another, for example, N_2. New polymers whose permeability and selectivity are higher than those of current membrane materials are being developed via synthesis of novel structures that prevent dense molecular packing, thus yielding high permeability, while restraining chain motions that decrease selectivity.

Pervaporation is a process in which a liquid is fed to a membrane process and a vapor is removed. The difference in composition between the two streams is governed by permeation kinetics rather than by vapor-liquid equilibrium as in simple evaporation. Thus, pervaporation is useful for breaking azeotropes and is

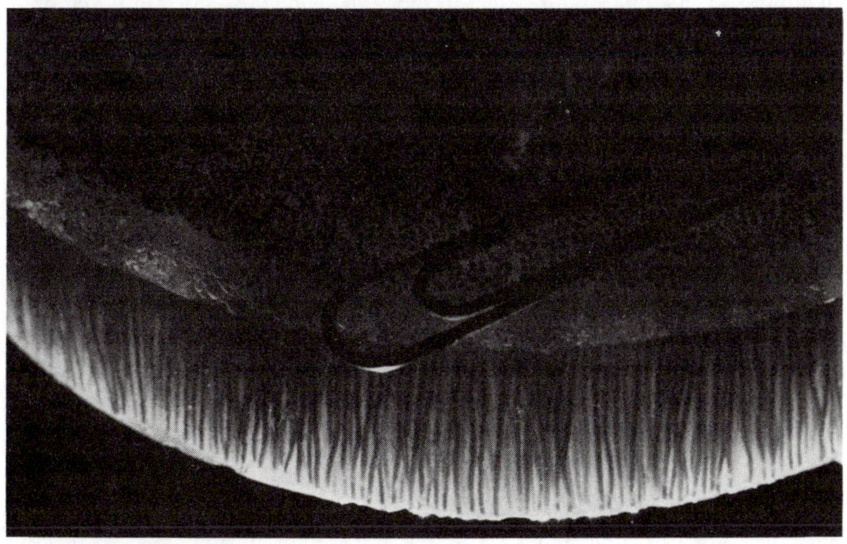

FIGURE 3.8 An industrial installation (top) of a membrane separator used in conjunction with a pressure-swing adsorption system to recover 90 percent of the hydrogen gas that normally would be lost in a waste stream. Also shown (bottom) is a cut-off section of a bundle of thousands of tiny hollow fibers made of polysulfone embedded in an epoxy tube sheet that fits into each tubular module shown. SOURCE: Photographs courtesy of Permea Inc., St. Louis.

generally used for alcohol-water separations, which become more important as biotechnology processes come into practice. Europe and Japan seems to be the leaders of research in this field. Major breakthroughs in membrane materials and fabrication are needed and appear to be possible.

Coatings

The U.S. coatings (paint) industry uses about 2.6×10^9 pounds of polymers per year and converts these to coatings having about $11B in sales. These coatings are applied onto goods valued at about $1T to $2T. For example, the typical cost of the paint on a house is about $500; on a car, $200; and on a refrigerator, $50.

Until the early 1970s most coatings contained only 15 to 30 percent paint solids, the remaining 70 to 85 percent being organic solvents, which were released as air pollutants when the films dried. Since then, reduction of solvent emissions has been the most important single driving force for technology change. Two kinds of 100 percent solids coatings are now being sold for specialized applications. Powder coatings are electrostatically applied and subsequently heat fused. Radiation-cured coatings are based on solventless liquid oligomers responding to ultraviolet or electronbeam cure. Solvent-borne high-solids coatings play an ever-increasing role. These are often based on oligomers containing hydroxyl groups as cross-linkable sites. Cross-linking is accomplished by formaldehyde-based methylolated or alkoxy-methylolated nitrogenous (amino) compounds. Acid-catalyzed heat cure causes the formation of multiple ether-based cross-links. Alternately, hydroxyl functional oligomers are cross-linked with isocyanates to form urethanes. Other common high-solids coatings have drying oil functionality and are cross-linked by air oxidation. Such high-solids coatings now contain about 20 to 50 percent volatile organic compounds (VOCs). The VOCs include solvents, by-products of the cross-linking reaction, and amines used to block catalysts. The role of aqueous coatings in reducing emissions of VOCs has increased greatly in the past few years. These coatings are based mostly on high-molecular-weight latex polymers and still contain some organic solvents to help film formation and wetting. Alternately, aqueous coatings are based on lower-molecular-weight hydrophilic polymers, which, unlike latexes, are synthesized not in water but in organic solvents and are subsequently dispersed into water. The ratio of VOCs to paint solids in aqueous coatings is now about 1:1 to 2:1. The environmental concerns leading to a reduction in VOCs were met by newly developed coatings based on new polymer chemistry that often provide better performance than the traditional coatings. In particular, resistance to weathering and to corrosive environments (including acid rain) was improved, not only for heat-cured but also for ambient-cured coatings. Lowering the cure temperature has allowed the application of coatings to a variety of structural plastics, which are starting to replace metals in many products.

The structure-property relationships for the great variety of new polymers are poorly understood. In fact, the fast empirical development of the new coatings has outstripped our scientific understanding. A few examples illustrate this. The most sophisticated coating resin system cannot be described solely by the overall monomer composition. Latexes are now being synthesized so that the monomer composition varies stepwise or gradually from the center to the surface of the particles. For example, a high-T_g (glass transition temperature) core and a low-T_g skin in the latex particles provide relatively hard coatings (related to high T_g) with ease of film formation (related to low T_g). Because the latex particles are very small, typically 50 to 300 nanometers (nm), it is practical to blend into them thermodynamically incompatible polymers. Layering, as described above, is one example. Acrylic monomers can be polymerized into urethane latexes, leading to separation into microphases within the latex particles or to formation of interpenetrating polymer networks. Alternately, latexes of differing compositions are synthesized separately and then blended. Also, latexes can be blended with separately synthesized low-molecular-weight water-soluble polymers. A single coating can be based on several polymers varying greatly in composition: polar and nonpolar vinyl polymers, urethanes, epoxies, polyesters, and alkyds. This way, the rheology, wetting properties, reactivity, and chemical resistance of the coatings can be optimized beyond what can be obtained by single polymer systems.

Apart from reduction of VOCs, another driving force for change is elimination of toxic ingredients from coatings. Some of the several sources of toxicity are cross-linkers for hydroxy-functional polymers: isocyanates or formaldehyde in amino cross-linkers. New cross-linking mechanisms are being sought. Promising early results involve reactions of carboxyl groups with epoxies or with nitrogenous heterocycles. The Michael reaction of acrylic double bonds with amines or activated methylene groups looks promising. Ketones or α,β-diketones are reacted with amines or carbohydrazides. Silane-functional polymers cross-link when exposed to atmospheric moisture. Some of these new systems provide chemically resistant cross-links even at ambient temperature, a great advantage for applications where high-temperature cure is not possible, such as car refinishes, aviation, implements, plastics, wood furniture, and architecture. New, very inert cross-linked resins in the coatings often provide good corrosion protection, allowing the use of nontoxic chrome-free corrosion inhibitors. Also, some new polymers provide improved interaction with difficult-to-disperse pigments so that toxic lead- or chrome-based pigments can be conveniently replaced by all-organic pigments. The greater variety of resins now available for coatings allows the elimination of toxic aromatic and ethylene-oxide-based solvents from high-solids or aqueous coatings.

Because the United States began to impose legislation to control the environment earlier than other countries, the U.S. coating industry gained early international technical leadership in nonpolluting coatings. These new technologies are now internationally available. However, the potential for further reducing VOCs and toxic ingredients and for further improving coating performance is huge.

For further progress, better understanding of solid- and liquid-state rheology and thermodynamics of complex resin systems is required. New chemical approaches for precision synthesis and for cross-linking of polymers and oligomers are also needed. Aqueous polymer systems are particularly fertile ground for new research. The preparation of complex acrylic latexes, urethane latexes, epoxy dispersions, solubilized polyesters, alkyds, and vinyl resins is still an empirical art.

Inorganic Polymers

The materials described above are made up primarily of polymer molecules based on covalently bonded chains in which carbon is the principal element. Polyethylene, polystyrene, poly(methyl methacrylate), and poly(vinyl chloride) are all based on carbon chains with differing side groups. Nitrogen and oxygen can also be incorporated into polymer chains, as in polyamides (nylons) and polyesters, but carbon atoms separate the hetero-atoms. Considering the importance and variety of organic polymers, natural and synthetic, it is remarkable that the chemistry of carbon is so unique and so dominant.

Other polymers exist in which carbon is less dominant. Polypeptides and proteins have one nitrogen for every two carbons, and the great variety is derived entirely from differences in groups attached to the alpha carbon atom. All of these materials are, however, entirely "organic" in character. Polyoxymethylene is a polymer with alternating oxygen and carbon along the main chain. It is a partially crystalline molding compound. Poly(ethylene oxide), with two carbons for each oxygen in the chain, has been of particular interest because of its water solubility.

Chains containing no carbon (C) exist, and virtually limitless compositional and structural diversity is accessible through utilization of more of the periodic table. Silicon, which is in the same group as carbon in the periodic table, is one such example. The vignette "Silicones" explores the properties and uses. At this time, however, the polysiloxane family is by far the most important. These materials have been available for several decades in the form of liquids, gels, greases, and elastomers that exhibit good stability and properties. They are the most thoroughly studied and highly commercialized class of inorganic polymers. Although the chain is entirely inorganic—with alternating single silicon (Si) and oxygen (O) atoms—organic side groups (usually methyl or phenyl) are attached to the silicon atoms. Many applications of polysiloxanes derive from the extraordinary flexibility of the siloxane backbone. The Si–O bond is significantly

SILICONES

Beginning in 1942, B-17 bombers were flown from factories in the United States to airbases in Great Britain. But before the Flying Fortresses could cross the Atlantic—much less raid German factories—a critical problem had to be solved. The thin air at cruising altitudes can be ionized by the high-voltage electricity of an aircraft ignition system. The effect is not large in the relatively dry air over land, but the water vapor in damp oceanic air is highly susceptible. The moist air would work its way into tiny pores in the rubber wires insulating the ignition system. Soon the pilot would notice a blue glow around the leading edge of an engine nacelle—a corona of electricity arcing from spark plug to cylinder, shorting out the plug. The engine would start misfiring, eventually dying altogether as more pistons quit. If more than one engine went south, the crew might have to ditch, and the "Fort" would be lost at sea. A silicone polymer developed for waterproofing electrical equipment aboard submarines proved to be the answer. Applied liberally to the spark plug wires and boots, the silicone grease kept the wires dry and the bombers airborne—it was, in fact applied to all U.S. bombers throughout the war.

Silicones are polymers whose backbones are long, flexible chains of alternating silicon and oxygen atoms. Dangling from the backbone like charms from a bracelet are side chains, usually small, carbon-based units, and the choice of these side chains gives silicones a remarkable range of properties. The water-repellent grease has oily, nonpolar side chains such as methyl groups. The nonpolar side chains and the polar water molecules do not mix, repelling the water from the silicone.

Another application of silicones depends on a careful balance between polar and nonpolar side chains. Small amounts of silicone foaming agents control the bubble size in polyurethane foams. A high proportion of polar side chains makes the foam foamier. The bubbles become bigger, forming open pores and producing the soft foams found in car seats and furniture cushions. Reduce the number of polar side chains, and the bubbles remain small. These tiny bubbles do not open up to form pores, and the foam is a much stiffer solid used for insulation.

Silicones have other, seemingly contradictory properties. A silicone resin coating the bread pans in a bakery keeps fresh-baked bread from sticking in the pan, and a liquid silicone polymer on the molds in tire factories does the same thing for newly made tires. But adding a "tackifying resin" makes the silicone sticky and produces the drug-permeable contact adhesive used on those skin patches containing nicotine (for smokers who are trying to quit) or scopolamine (for seasickness sufferers who are trying not to lose it).

Silicon and oxygen are the two most abundant elements on Earth, and they combine naturally to form silicates, including glass and such minerals as quartz and granite. These two elements were first combined synthetically—as silicones—in the United States in the 1930s. They were originally expensive and unhandy to make, but the discovery of a cheaper, easier method of producing them, coinciding with the interest in their novel properties sparked by World War II, started an avalanche of research into new uses for these versatile polymers that continues unabated today.

longer than the C–C bond, the oxygen atoms are unencumbered by side groups, and the Si–O–Si bond angle of 143 degrees is much more open than the usual tetrahedral angle of 110 degrees. These combined structural features increase the dynamic and equilibrium flexibility of the chain, causing, for example, the glass transition temperature of polydimethylsiloxane to be –125°C. Important applications include high-performance elastomers, membranes, electrical insulators, water-repellent sealants, adhesives, protective coatings, and hydraulic, heat-transfer, and dielectric fluids. The polysiloxanes also exhibit high oxygen permeability and good chemical inertness, which lead to a number of medical applications, such as soft contact lenses, artificial skin, drug delivery systems, and various prostheses.

Another family of inorganic polymers—the polyphosphazenes—is based on a chain of alternating phosphorus (P) and nitrogen (N) atoms. Over 300 different polymers have been synthesized, mainly by variation of the pendant groups. The pendant groups may be organic, inorganic, or organometallic ligands. The nature of the pendant groups affects the skeletal flexibility, solubility, refractive index, chemical stability, hydrophobicity, electronic conductivity, nonlinear optical activity, and biological behavior. Thus by choice of the appropriate side group, polyphosphazenes can be tailored for a variety of applications. These materials are prepared by methods that give little control over stereoregularity and, hence, mixed-substituent polyphosphazenes are amorphous. Glass transition temperatures as low as –100°C have been achieved, and elastomeric performance over a wide temperature range is characteristic of this family. Fluoroalkoxy substituents yield hydrocarbon-resistant materials that could be useful as fuel lines, o-rings, and gaskets in demanding environments. Ether side groups coordinate lithiumions, which leads to possible applications as polymeric electrolytes for high-technology batteries. The ease of side group substitution has also led to new applications in biomedical materials. Hydrophobic polymers such as poly[bis(trifluoroethoxy) phosphazene] minimize the "foreign body" interactions that normally occur when nonliving materials are implanted in contact with living tissues, such as blood. Hydrophobic polyphosphazenes are therefore good candidates for use in cardiovascular replacements or as coatings for pacemakers and other implantable devices. Hydrophilic or mixtures of hydrophilic and hydrophobic groups can be substituted to produce hydrophilic or amphiphilic polymers deliberately designed to stimulate tissue adhesion or infiltration or to generate a biochemical response. Unfortunately, while polyphosphazenes are an interesting class of materials that have physical and chemical characteristics that suggest many applications, they are costly to produce, and commercial success has consequently been modest.

Polysilanes (also called polysilylenes) have been the subject of research interest within the last decade. These all-silicon chains, with alkyl or phenyl side groups, are analogous to vinyl polymers, but they are made from silyldichlorides rather than from the analogue of ethylene. Linear and cross-linked structures can be made, leading to materials that are glassy, elastomeric, or partially crystalline. The chains of silicon atoms are flexible. Applications that have been proposed for polysilanes include ceramic precursors (for silicon carbide), photoresists, photoinitiators, and nonlinear optical materials. The use of these inorganic polymers as ceramic precursors is important because the precursor can be spun into a fiber that yields fibrous ceramics following processing. In fact, polysilanes are the basis of a process commercialized by Nippon Carbon Company to produce continuous NICALON® silicon carbide fibers for use in such applications as fiber-reinforced ceramic composites. Poly(dimethyl silane) or the cyclic oligomer are heated to 450°C to produce a polycarbosilane that can be spun into fibers. The fibers are heated first in air to create a silica skin that prevents melting of the fiber during subsequent pyrolysis in N_2 to produce silicon carbide (SiC).

The polyphosphates are inorganic polymers of interest in their own right, but their most important role is that they serve as part of the repeat unit in polynucleotides. The phosphate sequence is, therefore, being extensively studied with regard to biopolymer functions, for example in replication. Polysilazanes, with alternating silicon and nitrogen in the main chain, have also been prepared. Although these materials have not been as extensively studied as the polysilanes, it has been demonstrated that they can be used as precursors for silicon nitride ceramics.

Hybrid systems that combine organic ligands, polymers, or networks with inorganic polymers or networks, often referred to as CERAMERS or ORMOCERS (organically modified ceramics), represent some of the most structurally diverse and compositionally variable polymeric materials. The organic substituents can be designed as pendant groups that modify the inorganic polymer or network as in poly(organo siloxanes), as oligomers or polymers covalently bonded to the inorganic network, as interpenetrating networks, or as combinations of two or more of these functions. Hybrid systems are most often synthesized by low-temperature sol-gel hydrolysis using combinations of metal alkoxides and alkyl substituted and organofunctional alkoxysilanes: $M(OR)_n$ + $R_2Si(OR)_2$ + $YR'Si(OR)_3$, where R is an alkyl group, R′ is an alkylene, and Y is an organofunctional group such as vinyl, epoxy, or methacrylate.

Hybrid systems can potentially combine the advantageous properties of both organic and inorganic polymers. For example, epoxy-derived titanosiloxane polymers can be deposited at room temperature and cured at 90°C to produce optical quality, abrasion-resistant thin films as protective coatings on eyeglass lenses (and other optical elements) constructed of plastics such as poly(methyl methacrylate). The organic network provides resiliency and toughness, while the inorganic network provides hardness. The incorporation of elastomers such as polydimethylsiloxane and poly(tetraethylene oxide) into inorganic networks imparts flexibility to otherwise brittle materials, allowing large complex shapes to be processed rapidly without cracking. Independent free-radical and hydrolysis condensation reactions have been used to design hybrid gels that do not shrink upon drying.

The isolation of organic dye molecules, liquid crystals, or biologically active species in inorganic or hybrid matrices has led to a vast array of composite optical materials currently being developed as lasers, sensors, displays, photochromic switches, and nonlinear optical devices. These materials are superior to organic matrix composites because the inorganic matrix (normally silica) exhibits greater transmittance and is less susceptible to photodegradation. Organic molecules embedded in inorganic matrices can also serve as templates for the creation of porosity. Removal of the templates by thermolysis, photolysis, or hydrolysis creates pores with well-defined sizes and shapes. Inorganic materials with tailored porosities are currently of interest for membranes, sensors, catalysts, and chromatography.

Inorganic, organometallic, and hybrid polymers and networks represent a potentially huge class of materials with virtually unlimited synthesis and processing challenges. It is envisioned that future research will continue to explore the periodic table in search of new combinations of materials, new molecular structures, and improved properties. Hybrid systems appear especially rich for research in the area of multifunctional materials, that is, smart materials that perform several optical, chemical, electronic, or physical functions simultaneously. The development of hybrid materials that exhibit some of the extraordinary strengths and fracture toughness of natural materials such as shell and bone is also anticipated. The remarkable versatility of polyphosphazenes and polysiloxanes will continue to be exploited for biomedical applications such as drug delivery and organ and soft tissue replacement as well as advanced elastomers, coatings, and membranes.

The future of preceramic polymers and sol-gel systems appears bright. A major challenge is to develop synthetic routes to pure, stoichiometric nonoxide ceramics, especially SiC, that exhibit spinnability and high ceramic yields. New synthetic routes such as "molecular building block" approaches to multicomponent ceramics will be explored to prepare superconductor, ferroelectric, nonlinear optical, and ionic conducting phases, primarily in thin film form. The use of sol-gel processing to prepare "tailored" porous materials for applications in sensors, membranes, catalysts, adsorbents, and chromatography is an especially attractive area of research and development.

POLYMER PROCESSING

The growth of polymers both in volume and in number of uses, as described above, is in part related to their ease in processing. Contrary to popular perception, plastics are often more expensive than steel—that is, on a per-pound basis—but they are also much lighter than steel, glass, or aluminum. The great advantage to polymers lies in the many ways that they can be processed for functional uses such as coatings for surface protection, films for a wide variety of uses, fibers for fabrics and carpeting, and an enormous variety of molded shapes.

Melt Processing

Melt processing is the most widely used and generally the preferred processing method. It is used for polymers that become liquid at elevated temperatures so that they can be extruded into fibers, films, tubes, or other linear shapes or molded into parts of complex shape. Such processes involve much more than simply changing the physical shape of the polymer; they also influence phase morphology, molecular conformations, and so on and ultimately have an important role in the performance of the product.

Molding

A mold is a hollow form that imparts to the material its final shape in the finished article. The term "molding" is employed for processes involving thermosets and thermoplastics and includes injection, transfer, compression, and blow molding. The injection molding process is the most common method of making plastic parts. In that process, thermoplastic pellets are melted and pumped toward a melt reservoir by a rotating screw. When enough molten plastic has accumulated, the screw plunges forward to push the melt into a steel mold. The plastic solidifies on cooling, and the mold is opened for removal of the part. Injection molding cycle times vary from a few seconds to minutes, depending on the plastic and the part size. Molding machines have become very sophisticated, and they are capable of turning out large numbers of molded articles with little or no operator attention. The heated plastic conforms intimately to the polished mold surface, which may be of complex shape, and the part produced usually requires little or no further machining or polishing. The mold and the machine that delivers plastic to the mold can be quite costly; therefore, the technology is suited only to parts needed in large numbers. Even so, injection molding is a process capable of exceptionally low cost in comparison with production processes for metal or ceramic parts.

Some of the current challenges in polymer processing include developing new materials, achieving greater precision, pursuing process modeling and development, and recycling. Some examples of new materials are special blends of existing polymers, polymer composites with fiber reinforcement, and liquid crystalline polymers. Some of these new materials are expensive and may be difficult to form into desired shapes; however, they are of value to the defense and aerospace industries in applications in which weight and performance are more important than cost and processibility. In contrast, automotive and appliance industries use materials that are less expensive, readily molded, and dimensionally accurate. Yet there are limitations to what can be done, even with common materials. Molten plastics are viscous, and making thin parts may require high pressures. Further, the plastic shrinks as it cools, and this tendency must be compensated for by using oversized mold cavities. Molds are expensive, from several thousands to millions of dollars each.

Plastic materials are rheologically complex, and as a result many factors can affect the properties and dimensional accuracy of parts made from them. There are variations in operation of the molding machine, small temperature fluctuations, and differences in molecular orientation caused by flow into the mold. However, injection molding has been brought to levels that allow tolerances on small parts in the micrometer range. Among the high-performance plastics that have been introduced to meet the demands of the high-precision market are the thermotropic liquid crystalline polymers and low-viscosity versions of high-temperature materials such as polyetherimides and polyaryl sulfones.

Advances in processing are occurring at a rapid rate as on-line sensing, computing, and process feedback allow control and optimization of the molding process that were undreamed of only a few years ago. Parameters of importance include injection speed, peak pressure, hold pressure, and mold temperature, along with less obvious factors such as "cushion length" and position- or pressure-dependent cutoff. As the processing industry learns to take advantage of the capabilities of the new machines and materials, precision injection molding can be expected to make further inroads into the domain of machined metal parts.

Process dynamics and the properties of the finished article are critically dependent on the conditions of flow and solidification, down to the molecular level. As the mold is filling, the molten polymer solidifies first along the walls. The material that is farthest from the wall flows more rapidly, leading to a shearing and molecular elongation in the wall area. After the flow front has passed and the mold is full, the central regions solidify under conditions in which shear elongation is not a major factor. This solidification process leads to a morphology characterized by a skin of highly oriented polymer around a core of less oriented material. The two layers are mechanically and optically distinct. Control of these components through polymer composition and processing technology is a central issue in the production of precision, high-performance parts. Skin effects are most obvious in parts made from polymers that crystallize. Amorphous polymers are much less influenced. These morphological factors have important consequences for the failure mode and fracture mechanics of the finished part.

Additional processing techniques, such as gas-assisted molding and injection-compression molding, are gaining industrial acceptance. Materials suppliers are developing new plastics with enhanced flow characteristics and better physical properties. Compact disks can be made because of high flow grades of polycarbonate and the injection-compression molding process.

The importance of injection molding, and of precision injection molding in particular, can hardly be overstated. The economies that can be realized in the production of mechanically complex parts can contribute to the feasibility of large-scale manufactured products. For example, communications systems of the future, such as fiber optics to the home providing broad-band information, depend on many manufactured details that must be reduced in cost if the concepts are to succeed. Polymer-based solutions are essential to the realization of the promise of progress in diverse areas.

Extensive progress has also been made recently in blow molding, especially to form bottles for various packaging applications. Special grades of polymers with uniquely tailored rheological properties, via broad molecular weight distribution and chain branching, have been developed for this market. Stretch blow molding processes allow control of the development of chain orientation and crystalline structure for materials such as poly(ethylene terephthalate) to gain better barrier properties.

Extrusion

Melt extrusion processes are usually the most convenient, economical, and environmentally favorable for film and sheet manufacturing. Screw extruders, in which a rotating screw transports material through a heated barrel and a shape-forming die, are the heart of such processes. Extruder screws, which can be very sophisticated, are designed with the help of extensive computer-assisted modeling. Frequently, mixing, compounding, and devolatilization are also involved to process formulations that include special additives, such as antioxidants, plasticizers, flame retardants, lubricants, pigments, fillers, and other polymers. Films are formed through film blowing of thin-walled tubes or drawing and tentering of cast films. Optimization of the process requires fundamental understanding of material properties and processing characteristics.

The properties of the fabricated product are strongly dictated by the details of the fabrication process. Influential variables include uniaxial or biaxial orientation, degree of crystallinity, morphology of amorphous and/or crystalline regions, internal stress, and dimensional control. The complete system—from material design, synthesis, and formulation to product design and processing—must be addressed to achieve optimal material selection for a specific application. All elements of the system are important. Materials scientists and engineers, who understand structure-property relationships and can manipulate molecular design, work closely with process engineers and product designers in a systems approach to meet the growing demand for extrudable polymers that have specific characteristics.

Solid-State Forming

Solid-state forming is one of the newest polymer processing methods and is especially useful for increasing strength and modulus of polymeric materials. The latter involves achieving morphologies with well-aligned, extended, and closely packed chains, which can be done by synthesizing rigid rodlike polymers containing parasubstituted aromatic structures in the chain backbone or by processing conventional flexible chain polymers in ways that lead to similar results.

In this processing approach, a highly oriented and extended chain conformation with substantially increased tensile modulus may be achieved by solid-state deformation of thermoplastics using the extrusion or alternate drawing techniques, such as extrusion of supercooled melts, and by drawing from gel or dilute flowing solutions. High-density polyethylene (HDPE) has been widely studied, in order to understand the morphological transformations and because it can now be drawn into one of the highest specific tensile moduli and strengths in both fiber and film form.

Solution Processing

Not all polymers can be fabricated by the convenient and economical melt processing techniques; nor is this desirable in some instances. Certain polymers have such strong interchain bonding that they do not melt or flow when heated until they reach temperatures at which chemical degradation occurs. These intractable, nonmoldable polymers are often fabricated, typically into fibers or films, by solution methods. Some well-known examples include certain cellulosics and polyacrylonitrile materials, plus specialty polymers for high-performance fibers like Nomex®, Kevlar®, and polybenzimidazoles (see the vignette "Polymers Stronger Than Steel"). Often such materials are soluble only in aggressive solvents such as sulfuric acid. In the usual case, the polymer is dissolved to relatively high concentrations, and these solutions are extruded into the form of fibers or film, or solvent cast into films. The polymer is solidified by removal of the solvent by evaporation through optimized drying regimes (dry spinning) or by coagulation and extraction with a nonsolvent (wet spinning). Cellulose is not generally soluble without chemical modification, and so rayon fiber manufacture involves a series of reactions, first to make the polymer soluble and then to harden it by regeneration of the original cellulosic structure. These types of solution processes are less well understood than melt processing operations, owing to their added complexity and the relatively little attention paid to this area because of the small quantity of products produced in this way. Because of the necessary solvent recovery steps, these processes tend to be expensive. The products produced by them often have complex morphological structures associated with the manner in which the solvent is removed that can leave residual porosity or other structural features not typical of melt-fabricated materials. In

POLYMERS STRONGER THAN STEEL

One often thinks of polymers as soft plastic. However, when certain polymers are spun into fibers, the resulting materials are truly amazing. All polymers are long, chain-like molecules made up of many smaller molecules linked together. In flexible fibers, such as polyester, the chains are partially relaxed; that is, as we follow a polymer chain along a fiber, it passes alternately through regions that are well ordered and rigid and regions that are disordered and softer. However, the chains in these new rigid polymers are fully extended and lie parallel to each other, like a fistful of uncooked spaghetti. These fibers are unexpectedly strong for their weight. One such fiber, poly(paraphenylene terephthalamide), was found to have a tensile strength higher than that of a steel fiber of the same dimensions, yet it weighs one-fifth as much. The commercial development of this product was a long and very expensive process. More than 12 capital-intensive steps are required to convert the basic aromatic feedstocks into a strong polymer.

The key to realizing the outstanding inherent strength of these materials is the process by which the polymer is spun into fibers. The first attempts using traditional techniques involving injecting a stream of the polymer directly into a cooling bath did not result in an unusual material. However, it was found that raising the spinner head above the cooling water bath that the still-molten spun fiber falls into imparted unprecedented strength to the fiber. As the molten fiber falls toward the cooling bath, the polymer molecules become stretched and aligned. The oriented polymers can then form hydrogen bonds between molecules in adjoining chains, further strengthening the fiber. Because the molecules are fully aligned along their axis, forces on the fiber are absorbed by strong chemical bonds, not by the weak, loosely intertwined chains in the flexible polymers.

One of the aramid polymers, Kevlar®, has saved the lives of many soldiers and police officers. It has such high impact resistance that an aramid vest thin enough to be worn comfortably under the shirt will stop a bullet as effectively as steel plate. The helmets worn during Operation Desert Storm were also made of this remarkable material. This strong, lightweight material is now taking the place of steel belts in radial tires and is being exploited in a myriad of other applications that require a high strength-to-weight ratio or resistance to corrosive environments, such as the cables that anchor oil-drilling platforms at sea.

The molecular structure of a polymer is important, but, as shown above, the orientation of polymer molecules in relation to each other also plays a major role in determining the properties of the final product. The molecules' orientation can be greatly affected by the chemical engineers' choice of the way(s) in which the material is processed, allowing many more ways to customize polymers for their applications than are possible through chemistry alone.

addition, the potential solvent emissions and exposure of workers raise numerous environmental and health concerns. Nevertheless, such fabrication methods are the only option available for certain materials of critical importance. Interest in polymers with very high heat resistance and other special attributes is likely to increase the need for solution processing technology, because by the nature of their properties such materials cannot be melt fabricated.

In other cases, solution processing is required because of the nature of the product being fabricated rather than the nature of the polymer. Application of certain types of paint or coatings is most conveniently done by making the polymer fluid by dissolving it in a solvent and hardening by solvent removal. Melt fabrication cannot be used to economically perform certain operations like tablet coating (for controlled delivery) in the pharmaceutical industry. Formation of polymer membranes with complex structures consisting of extremely thin skins overlaying a porous support layer generally involves a variety of solution-processing protocols that can hardly be accomplished in any other way. Prepregs, used to make continuous-fiber-reinforced composites, often involve a solution processing step. "Molecular composites" based on rigid rod polymers dispersed in a matrix of a random coil chain polymer have attracted enthusiasm recently, and most techniques for forming them will invariably involve solution technology.

The state of scientific knowledge about the thermodynamics, rheology, diffusion, and morphology development in multicomponent polymer-solvent systems will need to be advanced in order to place these types of processes on a solid technical footing. For example, recent attempts to mathematically model the formation of asymmetric membranes have failed to give some of the insight critically needed simply because the fundamental elements of such models are not currently well-enough understood.

Environmental and health concerns associated with solution processing must be dealt with at several levels. There are opportunities for innovation in both the processes and the materials used. One avenue of interest is systems that use more benign solvents. Ongoing work in the area of supercritical fluid technology appears promising for some applications.

Dispersion Processing

Dispersion processing—the generation of particulate forms of polymers and their conversion into products—is a key technique that is likely to grow in importance, driven by both environmental and materials considerations. The polymer may be in the form of a dry powder, an aqueous dispersion (e.g., a latex), or a nonaqueous dispersion (NAD). A number of examples illustrate the current state of this technology and future directions.

Polytetrafluoroethylene, or Teflon®, is a powder after the polymerization process, and its melting point is so high that melt fabrication is usually not practical. However, by combinations of heat and pressure it is possible to fabricate solid forms in much the same way as used in powder metallurgy. Powder coatings are applied to various substrates by a number of techniques, including fluidized beds and flame heating. Powdered liquid crystalline polymers can be "extruded" below their melting points in much the same way as metal powders are formed. Technology is being developed to convert polymer powders into three-dimensional prototype parts by using a computer-driven laser-sintering technique. There is much to be learned about the physics of sintering polymer powders.

Emulsion polymerization is widely practiced because it is a convenient process for producing high-molecular-weight polymers using free radical mechanisms at high rates while controlling the exothermic nature of the reaction. In some cases, the resulting aqueous dispersion of the very fine polymer particles (called a latex) can be used directly for convenient fabrication of products. For example, the ubiquitous rubber gloves employed nowadays to protect against the spread of AIDS are made by dipping forms into a rubber latex followed by vulcanization. The popular water-based latex paints are made by emulsion polymerization. The paint film is formed by the evaporation of water, as opposed to organic solvents for oil-based paints. Painters also like cleaning up using water rather than paint thinner or solvents. The tiny emulsion polymer particles fuse into a continuous film driven by surface tension forces as the water evaporates. The rheological characteristics of the polymer must be designed such that fusion can occur without macroscopic flow of the coating. Environmental considerations strongly favor formation of coatings using latexes rather than solutions that "cure" by evaporation of organic solvents.

Nonaqueous dispersion technology has emerged as a means of applying high-performance coatings (e.g., automotive paints) while minimizing solvent emissions. These are sophisticated materials in which fine polymer particles are formed by dispersion polymerization in a nonaqueous environment in which dispersant polymer chains prevent coalescence of the particles through steric stabilization. By this route, high-solids fluids of acceptable viscosity can be rapidly applied to metal substrates to form high-performance coatings with reduced emission of organic solvents.

Inverse emulsion homo- or co-polymerization of water-soluble monomers such as acrylamide or acrylamide with co-monomers, such as cationic quaternary acrylates, represents a new commercially practiced method for the generation of materials that are suitable as flocculants. The materials as generated are extremely effective for purifying water, even at a level of 500 parts per million. When activated in water via modification of the surfactant or stabilizer, they can induce large flocs to be formed, which can be centrifuged from waste systems like sewage. The resulting flocs can be certrifuged, dried, and used as fertilizers. The resulting water is, of course, in a much purer state than before this treatment. Such operations are already in wide commercial use in many of the large cities of Europe and North America and are an example of how polymers can help clean the environment. Producing such polymers is already a multimillion dollar business.

Process Models

Extensive efforts have been devoted to the development of mathematical modeling or simulation of polymer processes to develop sophisticated manufacturing methods; however, much remains to be done. Most of these efforts have been applied to polymerization processes or to fabrication processes. The most essential ingredient of the former has been kinetic modeling of the reactions involved, while rheological behavior of the fluid polymer is usually the centerpiece of fabrication models. Each type may be coupled with mass and energy balances, along with accounting for heat, mass, and momentum transfer in varying levels of detail. Major issues in all cases are the level of sophistication of the models used and the accuracy of the input data. In the past, the sophistication or level of detail was limited by the computational power that could be brought to bear on the problem; however, this no longer seems to be the limitation. Future work in this area is expected to be extensive and more limited by the nature of the physical models available as the trend moves from simple process questions that can be answered by simple phenomenological models to complex product questions that require more detailed molecular models.

The most common driving force for developing process models is to aid in design. This function is often referred to as computer-aided design (CAD). For example, at the most elementary level a model of a polymerization process must be able to size the reactor and ancillary hardware and, in fact, to aid in making choices about the type of reactor system that would be most advantageous from a process point of view. Once a process configuration has been selected, computer-based models are generally able to predict temperature, pressure, and conversion profiles in time or space. More sophisticated versions are able to predict certain product qualities, such as molecular weight averages, copolymer composition, and branching frequencies. Models of this kind are valuable for modern computer-based process control schemes. They can also be of great value for evaluation of process safety issues. The future lies in going several steps beyond this to predict intrinsic polymer properties such as processing and end-use performance behavior. Such computer-based methods can be of value in the formulation, at least as a first estimate, and setting of process conditions to produce new grades of existing products.

Extensive efforts have been devoted to modeling fabrication operations such as extrusion, mixing, and molding operations. The sophistication employed depends on the objective to be achieved. Again, the most common motivation for model development is to aid design. For example, because molds are expensive to build, it is valuable to have models that are able to calculate whether a particular design will function as intended (e.g., fill up in the available time) for possible operating variables.

Most companies currently use externally or internally available software, combined with phenomenological rheological and heat transfer characterization of the polymer, to assess a proposed design before cutting metal. In spite of their computational complexity, such models answer only fairly simple process questions. Molecular-based models of flow behavior are emerging, which if properly combined with a process model are likely to be able to answer more subtle questions about both the process and the product. For example, in addition to simply determining whether the mold will fill properly, it would be useful to predict the crystallinity and molecular orientation (and thus the mechanical properties) of the product.

A good example of the progress being made in the development of molecular theories is in the field of rheology, which is one of the basic sciences needed for the understanding of polymer processing. Rheology involves macroscopic shape changes of a polymeric fluid in complex transient stress and temperature fields. In rheological studies, well-defined stresses or strains are applied in order to measure and to predict the mechanical behavior of complex materials. The most important recent advances in polymer rheology came about in the 1970s, when P.G. de Gennes proposed a model for macromolecular motion that was then explored by M. Doi and S.F. Edwards to derive a rheological constitutive equation for the stress in macromolecular fluids. Many rheological phenomena have since been explained in terms of molecular parameters such as molecular weight, branching, chain stiffness, surface interaction, and polydispersity. Many of the current efforts in rheology focus on the basic understanding of the effects of phase transitions and anisotropy. Other major efforts concentrate on new continuum models that expand and fine-tune the predictions of Doi and Edwards. These initiatives are supported by advances in experimental rheology that give direction to the development of its theory.

CONCLUSIONS

- Chemical products, in general, represent one of the few areas of the U.S. economy where the value of exports exceeds that of imports. Polymers contribute substantially to this positive balance of trade in chemicals. For example, plastics in both primary and nonprimary forms contributed $6.0B, or 37 percent, to the net positive trade balance in chemicals during 1992 (*Chemical & Engineering News*, 1993).
- Plastics manufacturing is an important part of the national economy. While the commodity polymers represent a maturing industry, significant process and product innovations continue to be introduced and appear likely to lead to healthy growth rates for the foreseeable future.
- A great deal of the current research and development on polymeric materials and associated processes is being driven by environmental considerations.
- Research on the recycling of polymers has been ongoing for more than two decades, and considerable progress has been made in some areas. A great deal of effort has been devoted to developing automated processes for segrega-

tion according to polymer type, and a uniform identification system has been agreed to by the industry to aid segregation by the consumer. Reprocessing of comingled waste leads to poor physical properties. Compatibilization may be technically possible, but it is quite expensive in relation to the value of the products produced. On the other hand, some plastic products appear to be easily recycled. For example, poly(ethylene terephthalate), which is used for soft drink bottles, is easily identified and is usually relatively free of contamination. A number of ways of reusing these materials have been identified, and some are in use. Perhaps the most significant are those that cause the polymer to revert chemically to oligomeric species that can be repolymerized into poly(ethylene terephthalate) or other polymeric products.

- While innovations in polymer recycling are needed, other options must also be pursued. Opportunities include source reduction and design of products with recycling in mind. The potential value of biodegradable polymers as part of the solution to the solid waste and litter problem needs to be better understood. Such materials are likely to be much more expensive than the relatively inexpensive polymers currently used, and their performance may not be as good. It is not clear that any materials biodegrade in landfills. In any case, the release of the low-molecular-weight degradation products into the environment could lead to more serious air or water contamination concerns. Incineration for fuel value is another, and perhaps the ultimate, form of recycling of polymers. Most polymers are derived from oil, and about 95 percent of all the oil produced is burned for its energy value; thus oil converted into polymers is simply being borrowed for a while to be used as a material prior to returning to its ultimate fate of being burned for its energy. Of course, concerns about the impact of incineration on health and the environment need to be resolved.

- Polymer interfaces are key to the performance of composites, blends or alloys, lubricants, adhesives, coatings, and thin films. Advances in the fundamental understanding of these interfaces and methods to engineer desired performance of these surfaces will no doubt lead to improved products and a competitive edge.

- New engineering plastics, including some blends or alloys, with ever-increasing performance characteristics continue to be introduced and in many cases are being used for structural applications traditionally dominated by other materials, mainly metals. Ease of fabrication into dimensionally precise parts with high-quality surface finish is one driving force.

- For polymer-based materials being used in highly critical structural applications, there is need for a better understanding of the mechanical, chemical, and environmental factors that affect their useful lifetimes and for methodologies to predict lifetimes in complex situations. Fracture mechanics techniques are not yet used to the same extent for polymers as for other structural materials.

- There are opportunities for new polymer systems with controlled permeability properties for use in food packaging, medicine, clothing, agriculture, industrial manufacturing, pollution control, water purification, and other applications. Innovative synthesis, knowledge of structure-property relations, and fabrication technologies will need to be integrated to achieve functional products that are cost-effective.

- Instrumentation is emerging that makes possible real-time determination of chemical and physical structures during processing, allowing instantaneous comparison with process simulation and control models and quality assurance.
- Rapid progress in the development of production processes is being made through implementation of a "systems approach" wherein the materials, their composition, and the process are all considered variables in an iterative process. The systems approach, in comparison with the traditional "compartmentalized" approach, facilitates rapid identification of the critical material and processing parameters and aids manufacturing procedures.
- The relationship between processing and properties, particularly for complex polymer systems such as blends, composites, and liquid crystalline polymers, is not fully developed. The role of rheology in the development of structure is at a particularly rudimentary stage, and it appears that significant advances could be possible. The development of real-time probes of structure during processing would be of immense value. Of course, a linkage must be made between the structure generated during processing and the performance of the product.

REFERENCES

Barbero., E., and H.V.F. Gangarao. 1991. "Structural Applications of Composites in Infrastructure." *SAMPE Journal* 27 (No. 6, Nov./Dec.):9.

Chemical & Engineering News. 1991. Vol. 69, No. 23, June 10, p. 39.

Chemical & Engineering News. 1992. "Facts & Figures for the Chemical Industry." Vol. 70, No. 26, June 29, pp. 62-63.

Chemical & Engineering News. 1993. "Facts & Figures for the Chemical Industry." Vol. 71, No. 26, June 28, pp. 38-83.

McDermott, J. 1993. *Advanced Composites: 1993 Blue Book*. Cleveland, Ohio: Advanstar Communications.

Modern Plastics. 1982. "Materials 1982: A Modern Plastics Special Report." Vol. 59, No. 1, January, pp. 55-87.

Modern Plastics. 1992. "Resins 1992: Supply Patterns Are Changing." Vol. 69, No. 1, January, pp. 53-96.

Pasztor, A. 1992. "Composites Makers Take Aim at Nondefense Markets; As Pentagon Budget Shrinks, Firms Seek to Dispel Cost, Safety Concerns." *Wall Street Journal*. August 26, pp. B3-B4.

Strathman, H. 1991. *Effective Industrial Membrane Processes: Benefits and Opportunities,* M.K. Turner, ed. New York: Elsevier.

4

Enabling Science

Enabling science describes the fundamental work that makes possible the development of technological applications. This work includes synthesis, characterization, and theory.

POLYMER SYNTHESIS

Synthesis provides the underpinnings for all advances in the science and engineering of polymeric materials. The past decade has produced many notable synthetic advances: the extension of living polymerization methods to new classes of reactive intermediates and new classes of monomers, the appearance of hyperbranched polymers, the controlled preparation of inorganic polymers and organic-inorganic hybrids, the development of efficient biological polymerization strategies, and many others.

This section considers the objectives of synthetic polymer chemistry, both fundamental and practical. The most important issues examined below include:

- Control of chain architecture,
- Synthesis of polymers of controlled end-group structure,
- Design and synthesis of polymers with outstanding thermal stability or useful electronic properties,
- Modification of polymer surfaces,
- Development of new polymerization methods,
- Challenges in reactive processing, and
- Interplay of organic, inorganic, and biological chemistry in producing novel macromolecular materials.

Some of the key synthetic advances of the last 10 years are noted, and unsolved problems and important research directions are identified.

Control of Chain Architecture

Topology

Chain topology controls many critical properties (e.g., crystallization, solubility, and rheological behavior) of polymeric systems. The development of linear polyethylene via the Ziegler and Phillips processes in the 1950s provides the most important example, wherein "straightening of the chain" resulted in a rise in T_m of ~30°C and significant improvements in strength and toughness. Control of topology in this case led directly to a new class of materials now sold in quantities of millions of tons per year.

Current synthetic methodology affords a high level of control of chain structure, and syntheses of linear and branched chains, stars, rings, and combs have been achieved. A striking recent advance has been the preparation of hyperbranched—or dendritic—macromolecules, in which iterative branching steps lead to structures in which segment density grows rapidly as one proceeds radially from the molecular "core" (Figure 4.1). Dendritic polymers of narrow molecular weight distribution have been prepared by laborious stepwise synthesis, while "one-pot" methods have been developed for polydisperse samples. In some instances, remarkable improvements in solubility (in comparison with linear analogues) have been reported for hyperbranched chains, and there has been plausible speculation that dendritic polymers may have useful sequestration and reactivity properties. Commercial applications for these materials have not emerged as of this writing.

Polymeric rotaxanes, in which the chain backbone is threaded through a series of macrocycles, have also been reported recently. In this arrangement, the macrocyclic "rotors" are not covalently attached to the "axle" but instead are constrained by bulky end groups from topological dethreading. It has been suggested that controlled, reversible switching of rotor positions on the axle might provide a basis for functional molecular devices (Figure 4.2).

Even given these advances, there is no shortage of intriguing and important topological issues yet to be addressed, and it appears likely that attention will shift to molecules of even greater complexity. Preliminary reports of two-dimensional ribbonlike polymers have appeared, as have descriptions of synthetic DNAs with the topological character of a cube. Particularly intriguing is the prospect of exploiting both covalent and non-covalent interactions, to provide control not only of topology, but also of the molecular geometry over large length scales in real space.

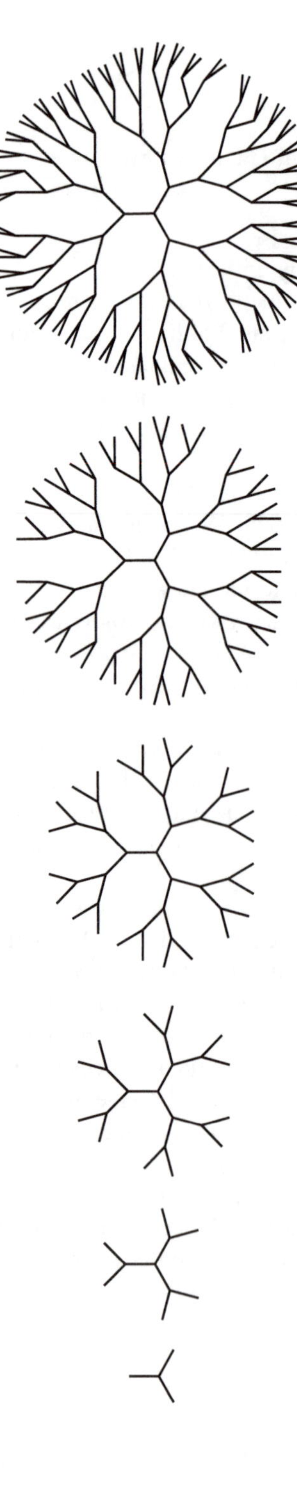

FIGURE 4.1 A schematic illustration of the divergent synthesis of dendritic polymers. Segment density grows rapidly as one proceeds radially from the molecular "core." SOURCE: Reprinted with permission from Tomalia et al. (1992). Copyright© 1992 by the American Chemical Society.

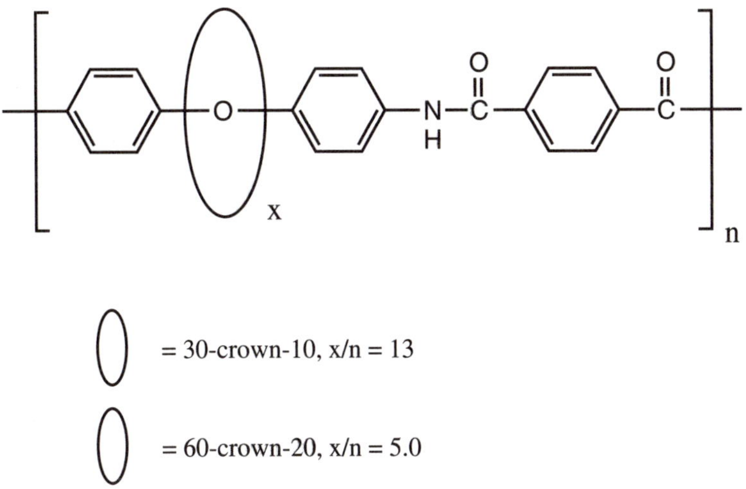

= 30-crown-10, x/n = 13

= 60-crown-20, x/n = 5.0

FIGURE 4.2 A polymeric rotaxane with an aromatic polyamide "axle" and macrocyclic polyether "rotors." SOURCE: Reprinted with permission from Gibson et al. (1992). Copyright© 1992 by the American Chemical Society.

Length

Many desirable properties of polymers, such as mechanical strength, thermal behavior, processibility, diffusion, miscibility, and adsorption at interfaces, are dependent to a large extent on the molecular weight. Hence, synthetic polymer chemists have directed attention to development of methods appropriate for molecular weight control. Great strides have been made, especially in the area of living polymerizations. Over the past decade, besides improvement in living anionic techniques, several new methods have been devised and/or improved. Notable among these are group transfer polymerization (GTP) for polymerization of (meth)acrylates (Figure 4.3), ring-opening metathesis polymerization for cyclic hydrocarbons (ROMP), Lewis base-mediated cationic polymerization for vinyl ethers, and immortal (metalloporphyrin-catalyzed) polymerization for heterocyclic monomers. All of these methods are capable of controlling not only the average chain length but also the molecular weight distribution.

However, chain length control is still statistical. One can argue that one of the current major challenges facing the polymer chemist is the preparation of polymers of absolutely uniform chain length for a wide range of polymers. There are indications that this may be met through the use of genetic techniques and solid-phase synthesis. How properties will be affected and which properties will be influenced by the truly monodisperse materials are two questions to be explored. Another challenge is in condensation polymerization, where chain length control is not precise. The use of the cyclic oligomer method recently developed shows promise, but there is a need for further studies. A thorough mechanistic

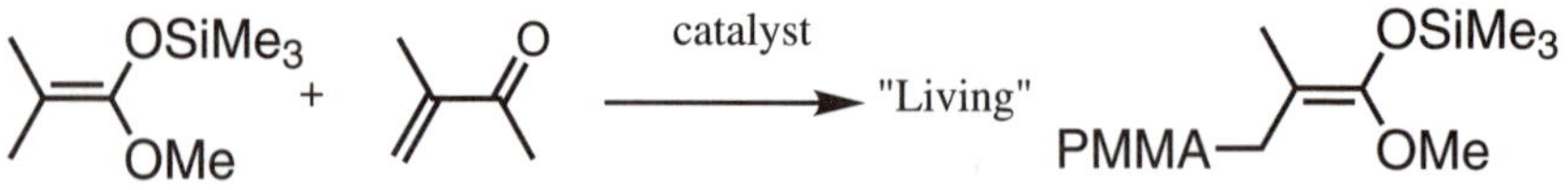

$$\text{Me}_2\text{C}{=}\text{C}(\text{OSiMe}_3)(\text{OMe}) + \text{CH}_2{=}\text{C}(\text{CH}_3)\text{C}(\text{=O})\text{OMe} \xrightarrow{\text{catalyst}} \text{"Living" PMMA}{-}\text{CH}_2\text{C}(\text{CH}_3){=}\text{C}(\text{OSiMe}_3)(\text{OMe})$$

$\bar{M}_n$ up to 250,000

$\bar{M}_w / \bar{M}_n < 1$

FIGURE 4.3 Group transfer polymerization (GTP).

study to delineate the factors affecting chain termination and chain transfer, events that adversely affect chain length control, is also needed if one is to solve completely the problem of molecular weight control.

Sequence

Control of monomer unit sequence in copolymers is the most difficult of the fundamental problems facing polymer synthesis. Whereas nature prepares macromolecules of precisely defined length and sequence, synthetic polymers are mixtures of chains characterized by substantial heterogeneity. As described in the preceding section, considerable progress has been made in the development of polymers with narrow distributions of chain length. By comparison, the control of sequence that can be exercised is primitive and is limited essentially to the preparation of statistical, alternating, and block copolymers.

Three approaches to the synthesis of copolymers of controlled sequence are available, but all suffer from significant limitations. Solid-phase methods, developed in the 1960s for peptide synthesis, remain slow and costly. Genetic engineering techniques are more rapid and efficient, but are for the foreseeable future restricted to the preparation of polypeptides. The third method, now in its infancy, consists of the controlled, portionwise addition of monomers to living chain ends. While this method is unlikely ever to lead to precise control of sequence at the monomer level, the preparation of complex architectures comprising many short monomer blocks appears to be feasible. One might expect such controlled-sequence copolymers to exhibit unique and interesting behavior in self-assembly, at interfaces, or in optical or electronic applications.

Isomerism

The geometrical, stereochemical, and regioisomeric structures of the constituent chains have major influences on the transition behavior, morphology, thermal stability, and so on, of polymeric materials. Geometrical isomers can be most easily illustrated by pointing out the possible microstructures available in polydienes, such as polyisoprene and polybutadiene. Microstructures of polybutadiene are shown in Figure 4.4.

Polybutadiene can exist in the *cis* or *trans* arrangements or in vinyl or 1,2-structures. Furthermore, the 1,2-microstructure can exist in highly isotactic or syndiotactic configurations or in a mixture of the two, which one describes normally as atactic. The importance of these various isomers varies widely. The *cis* structure has a remarkably low glass transition temperature and is widely used in tires and other important elastomeric applications. By contrast, the *trans* structure is highly crystalline and currently has no important applications, even though various biomedical uses (e.g., casts) have been proposed in the past. Both highly isotactic and syndiotactic semicrystalline 1,2-polybutadiene materials have been prepared by coordination catalysis; however, at this time neither has found significant application. Atactic 1,2-polybutadiene is accessible by anionic polymerization and has been considered as a potential hydrophobic, moderate-cost starting material for low dielectric thermosetting networks. However, significant applications of this material have not been developed.

The current state of the art in these areas utilizes principally free radical, anionic, and coordination catalysis. Free radical polymerization of butadiene in emulsion is practiced and produces a mixed *cis-trans* structure with about 20 percent of the 1,2-microstructure. The ratio of *cis* to *trans* units is polymerization temperature dependent. Current practices produce approximately 50 to 60 percent *trans,* ~20 percent 1,2-microstructure, and the remainder *cis* 1,4-microstructure. Anionic polymerization in hydrocarbon solvents is industrially important and produces 90 percent 1,4-microstructure with short sequences of *cis* and *trans* arrangements predominating. About 10 percent of the chains are atactic 1,2-units. This material is used in a variety of applications and displays a glass transition temperature of about –90°C. Coordination catalysis can generate rather high isomeric purity in all of these systems.

Regioisomerism can include variations on the usual head-to-tail enchainment that is observed in most chain polymerization of vinyl monomers. In general, both electronic and steric driving forces produce largely head-to-tail enchainment. However, it is well known that small amounts of the abnormal head-to-head and tail-to-tail structures are found in a variety of materials, including poly(vinylidene fluoride), poly(vinyl chloride), and to some extent, poly(vinyl acetate). Minor imperfections of this type can influence critical performance or processing parameters, such as piezoelectricity or thermal stability in the melt. This has been a problem in poly(vinylidene fluoride), where variation in polymerization conditions can result in a proportion of head-to-head structures as high as 20 to 30 percent. These structures are inevitably less thermally stable and, in general, are undesirable. Exceptions certainly are possible, particularly where controlled instability is desired, as in some resist materials.

Microstructures of polybutadiene

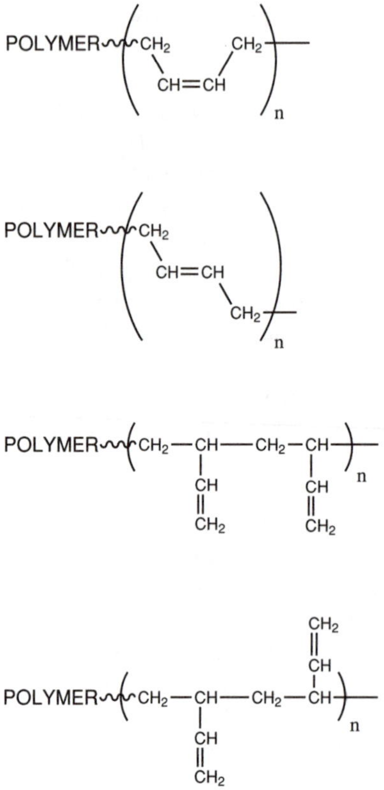

cis-1,4-polybutadiene
T_g = -100°C
Hardly any crystallinity: T_m = 0°C

trans-1,4-polybutadiene
T_g = 80°C
T_m can be up to 140°C

isotactic 1,2-polybutadiene
T_m = 125°C
High T_m is not very useful because the material is oxidatively unstable.

syndiotactic 1,2-polybutadiene
T_m = 156°C

atactic 1,2-polybutadiene
T_g = 0°C
Results from anionic polymerization techniques.

FIGURE 4.4 Microstructures of polybutadiene.

The tacticity of vinyl polymers is critical to the control of the crystallinity and thermal transition behavior of these materials. Free radical polymerization produces mostly atactic structures, with the temperature being one important parameter, which can modestly control placement. In general, ionic polymerization, and especially coordination catalysis, is required to produce either highly isotactic or highly syndiotactic materials. A limited number of possibilities exist with anionic polymerization, for example, with certain methacrylates and with cationic polymerization in, for example, vinyl alkyl ethers. Recent advances in this area concern the preparation of syndiotactic polystyrene with titanium-alu-

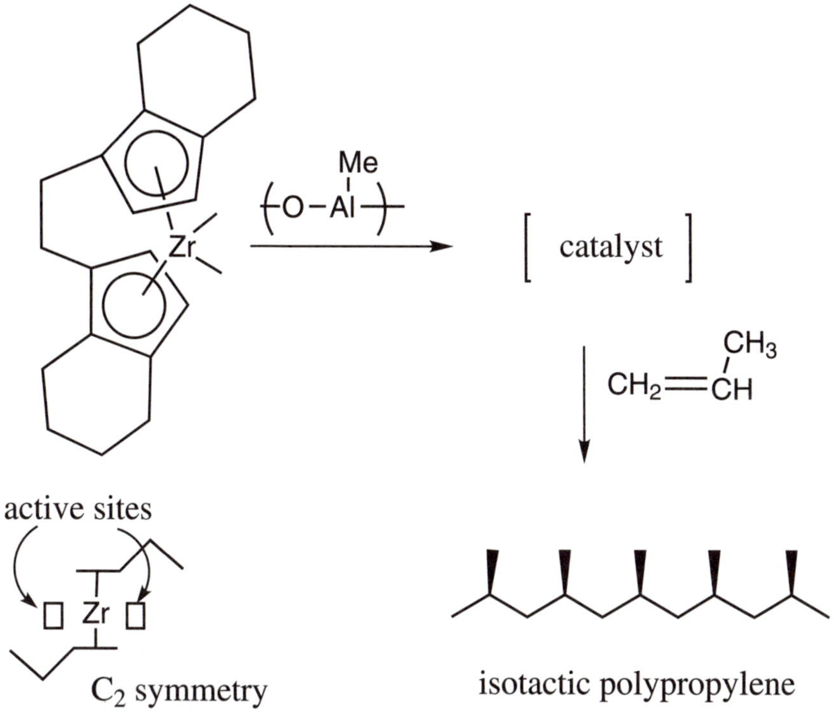

FIGURE 4.5 Chiral complex for the control of tacticity of polypropylene.

minum catalysts and the production of stereoblock polypropylenes with soluble metallocene catalysts (Figure 4.5).

Control of macromolecular asymmetry has also been achieved within the last decade, such that optically active materials can now be prepared on the basis of a controlled helix sense, that is, a predominance of left- or right-handed helices in the solid polymer. Such materials have found application as media for chromatographic resolution of enantiomers.

Advances in catalysis that could produce better understanding of polymerization mechanisms are needed to further refine the microstructure and to produce improved materials.

Synthesis of Polymers of Controlled End-Group Structure

An increasingly important class of polymers are telechelic polymers, which contain reactive end groups that can be used to further increase the molecular weight of a polymer during processing or to generate block copolymers. Perhaps the most widely used materials of this type are the polyols or polyglycols.

These nominally difunctional materials are used in a wide variety of applications in which they are reacted with isocyanates to prepare segmented polyurethanes and polyureas (Figure 4.6). Polyols can be prepared by either ring-opening or step-growth polymerization. In the latter case, the end-group functionality will be defined by the component that is present in molar excess, and the molecular

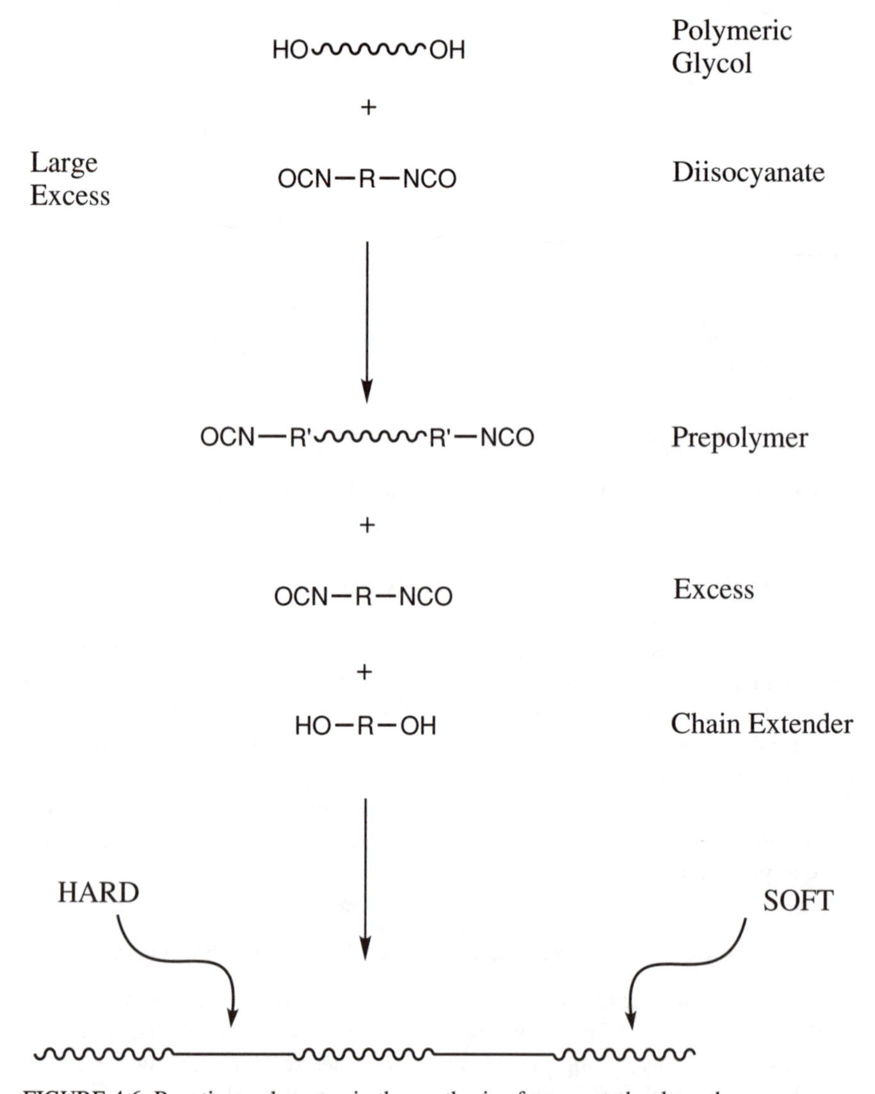

FIGURE 4.6 Reactive end groups in the synthesis of segmented polyurethanes.

weight can be controlled by the stoichiometry. Thus, the same chemistry that allows end-group control also can produce other, more sophisticated block or segmented copolymers. The success of the polyglycols, such as polypropylene glycol and polytetramethylene glycol and their copolymers, demonstrates the value of this approach. For example, urethane foams for automobile seats are generated from the polypropylene glycols. Thermoplastic polyurethanes, which are important in a variety of applications, are synthesized from the polyether systems, as well as polyester polyols.

Telechelic polymers can also be produced by living polymerization, and either a functionalized initiator or a terminator can be used to introduce the reactive groups of interest (Figure 4.7). Alternatively, a difunctional initiator can be used, and the reactive ends can be functionalized. Both approaches have been demonstrated for anionic and group transfer polymerization, and an analogous technology for cationic polymerizations is beginning to emerge.

In cases in which the reaction is not living and cannot be controlled to be living, chain transfer with a functionalized agent is the only solution for the preparation of telechelic polymers. Some success has been realized using this technique for radical polymerizations, and recent advances have been made in the preparation of telechelic polymers using difunctional acyclic olefins as chain transfer agents in ring-opening metathesis polymerization. This system is particularly suited to this approach owing to the near identity of the reactivity of the acyclic chain transfer agent and the cyclic olefin monomer.

In spite of this progress, telechelic polymers synthesized by chain processes remain difficult to prepare on a large scale and with a high degree of end-group functionality. New techniques and methods are essential to prepare such materials. For example, there are no straightforward routes for the preparation of telechelic polyethylene and polypropylene.

Major advances will come from the development of new techniques to control the molecular weight and molecular weight distribution of step growth polymers and the synthesis of chain growth polymers that have precise end-group structures.

Design and Synthesis of Thermally Stable Polymers

Thermal stability has been defined as the capacity of a material to retain useful properties for a required period of time under well-defined environmental conditions. Many factors contribute to heat resistance, including primary bond strength, secondary bonding forces (hydrogen bonding, dipolar interactions), and resonance stabilization (in aromatic structures). The mechanism of bond cleavage (particularly with respect to whether the broken chains can be combined or further unzipped) must also be considered. For example, α–methyl-substituted macromolecular materials often will regenerate sizable amounts of monomer, whereas highly aromatic condensation polymers and even vinyl polymers with

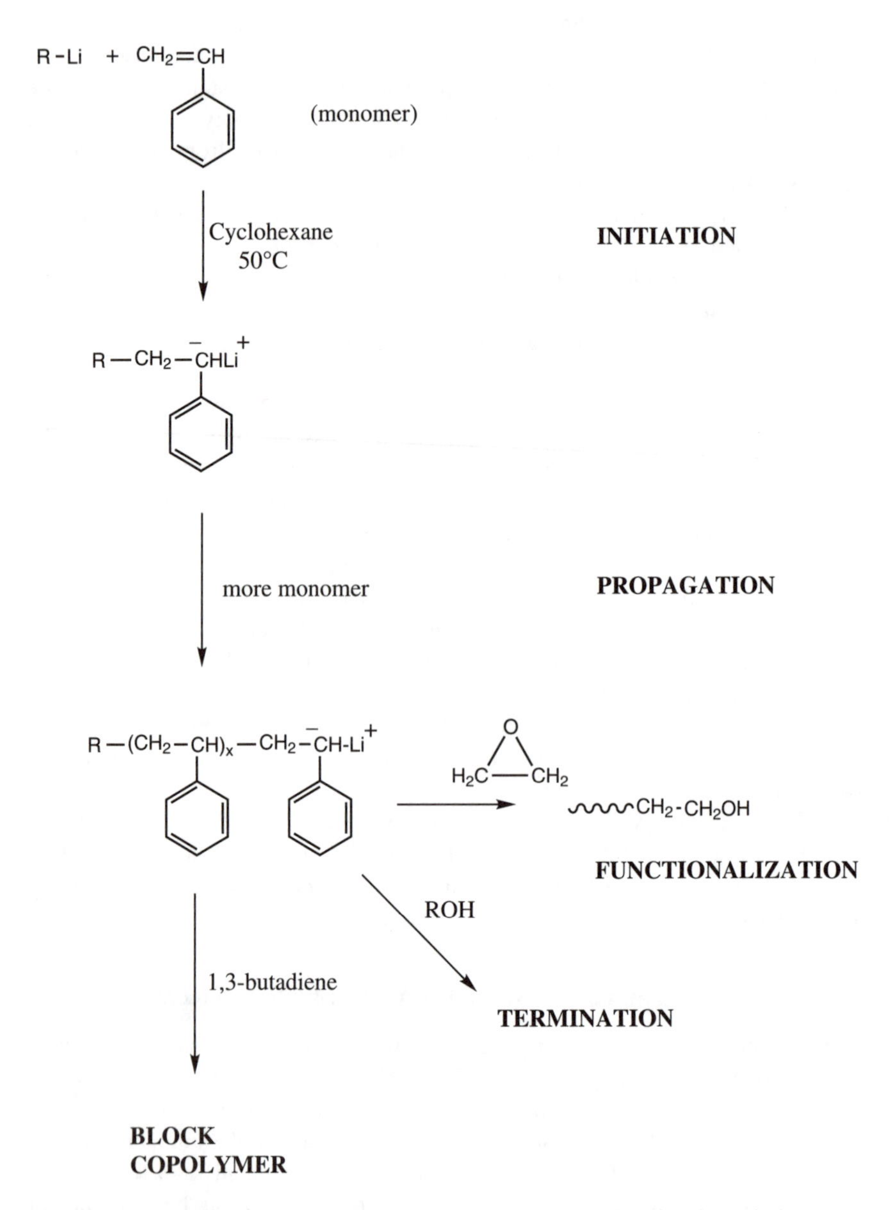

FIGURE 4.7 Introduction of reactive end groups by use of a functional terminator.

tertiary hydrogens will often prefer to recombine. Of course, molecular weight and molecular weight distribution can play a role, along with molecular symmetry. This latter parameter influences whether or not ordered morphologies, such as a semicrystalline structure, can be generated. The possibility of cross-linking, either randomly or via a terminal unit, to produce networks may be significant, along with the overall "purity" of the system. This last point includes the retention of catalyst fragments utilized for the polymerization. Concepts used in the molecular design of processible macromolecular materials with exceptional thermo-oxidative stability are summarized in Figure 4.8, where a number of the parameters described above are illustrated in the context of two of the major thermo-oxidatively stable materials systems, the poly(aryl imides) and the poly(aryl benzoxazoles).

Synthesis of Conjugated Polymers

Conjugated polymers show considerable promise in electrical and optical applications. However, the synthesis and use of these materials have presented a variety of formidable problems. In general, these materials by their very nature are difficult to process. The parent systems are rigid and flat and exhibit very strong interchain interactions. Consequently, they are insoluble and difficult to fabricate. Two major approaches have been used to allow these materials to be fabricated: (1) the development of precursors and (2) the use of solubilizing side groups.

Precursor routes to conjugated polymers can be as simple as the deprotonation of a heterocyclic polymer or as complex as a major carbon skeleton rearrangement. The fabrication of polyaniline is an example of the former, and the conversion of polybenzvalene to polyacetylene is an example of a major carbon backbone rearrangement.

Major advances have been made in solubilizing conjugated polymers by attaching side chains. For example, polythiophene is insoluble, whereas poly(n-hexylthiophene) is processible by spin casting and by the use of other solvent-based fabrication techniques. A similar approach has been used to solubilize poly(paraphenylene)s and poly(phenylene vinylene)s (PPVs; Figure 4.9).

Problems are associated with each approach. The precursor approach requires that chemistry be performed on the polymer. In many cases, this leaves defects and voids. The substituent approach results in defects in the polymer chains and diminishes interchain interactions that are important to some desired properties. Given the real and potential advances in the use of conjugated polymers in device applications, a vigorous synthetic attack on such problems is badly needed.

PARAMETER	EXAMPLES
• Chain Rigidity (Bond Strength)	Aryl Imides 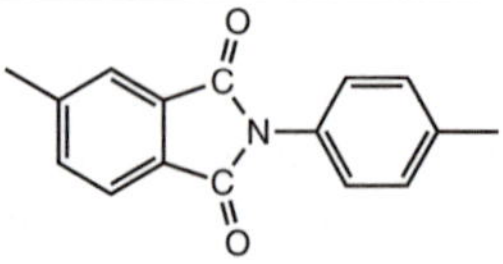 Aryl Benzoxazoles 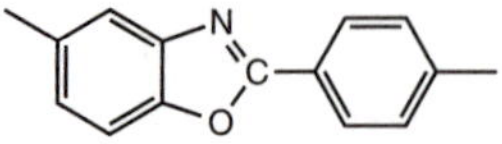
• Controlled Morphologies	Copolymers (Statistical or Block)
• Thermally Stable Heterocyclic/ Heteroatom Units	Imides, Benzoxazoles, Phenyl Phosphine Oxides, Fluorinated Materials
• Conformational Non-Planarity	Biphenyls 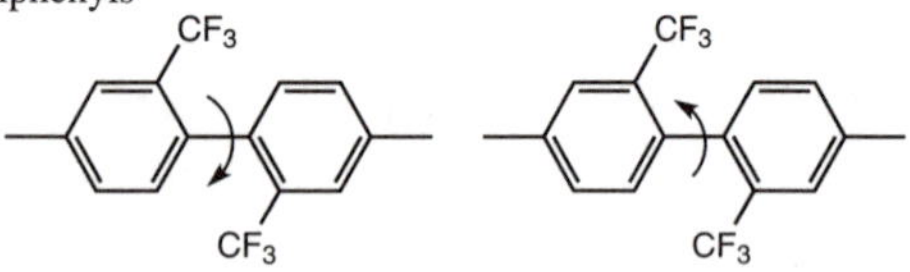 (CF_3, CF_3; CF_3, CF_3) Phenyl Phosphine Oxide 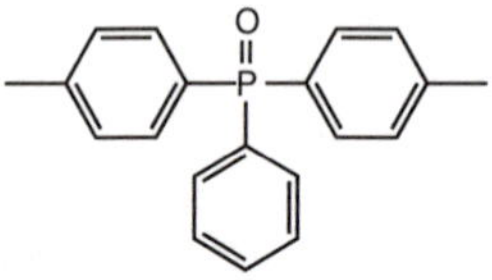 "3F" Systems 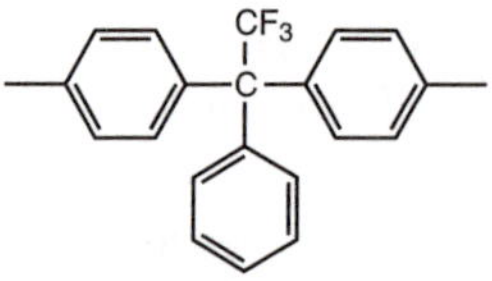 (CF_3)
• Chain Length/End Group Control	Non-Reactive Thermoplastics; Reactive, Uniform Network Thermosets

FIGURE 4.8 Concepts for molecular design of processible macromolecules with exceptional thermal and thermo-oxidative stability.

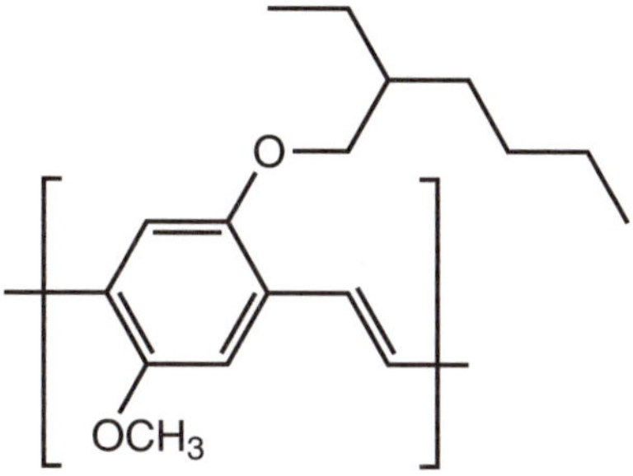

FIGURE 4.9 Poly(2-methoxy-5-(2-ethylhexyloxy)-1,4-phenylene vinylene), a soluble form of poly(phenylene vinylene).

Modification of Polymer Surfaces

Chemical surface modification of solid organic polymers is generally carried out to change chemical or physical surface properties of preformed objects (e.g., film, fibers, and bottles) when alteration of the polymer bulk properties is not desired. This type of modification alters only a small fraction of the sample material (typically the outermost 10 Å to 1 mm). This is in contrast to solid polymer modification reactions such as vulcanization of diene elastomers to prepare cross-linked rubbers or sulfonation of cross-linked polystyrene beads to prepare ion exchange resins, in which the modifications occur throughout the sample. Surface modifications proceed "on" polymers (normally at the solid-solution interface) as opposed to "in" polymers.

Surface modification of polymers is a topic of considerable current interest and will continue to advance technology and various research fields in the future. As a field, chemistry at polymer surfaces is becoming scientifically interwoven with the research fields and technologies of chemical, biological, and physical sensors, protective layers, adhesion, membranes, electrodes, information storage, and optical devices and will, in the future, have an impact in these areas.

The classical methods for surface modification (i.e., plasma modification, surface graft polymerization, and chemical reaction) are crude and were developed to treat high-volume (high-surface-area) commercial polymers. Plasma modification techniques are widespread and versatile. Although surface chemistry can be changed to effect the desired properties, the method is brutal, and cascades of reactions involving homolytic bond scission, fragmentation, and cross-linking occur. Little control is possible. Surface graft polymerization is also common, but the polymer-polymer interface structure and graft density and location have generally not been controllable. The classical chemical methods are likewise crude. Oxidations of hydrocarbon polymers, reductions of fluoropolymers, halogenations, dehydrohalogenations, and hydrolyses are important modification reactions but are most often corrosive and not controllable.

The difficulties of performing specific organic chemistry at the solution-solid interface have become appreciated over the past 10 years, and strategies to perform organic chemistry in a controlled fashion in this milieu have been developed. The structure of the solution-solid interface in a modification reaction is extremely system dependent, and a large variety of situations can be envisioned. The mobility of polymer chains in contact with the solution depends on the polymer morphology (degree of crystallinity) and structure (glass transition temperatures span 400°C for common polymers) as well as the solution—to what extent it wets or swells the solid. There is no generic picture of a solid polymer-solution interface. The fact that the interface structure changes upon reaction further complicates the possibilities. Surface modifications have been reported that are either "autoinhibitive" or corrosive. In the former type of modification the solution interacts with the product surface to a lesser extent than it does with the starting polymer, and the reaction stops when a thick layer of modified material forms and acts as a barrier to further reaction. In the latter type of modification the solution interacts more strongly with the product surface than it does with the starting polymer, and the reaction proceeds to a greater depth, or the product dissolves as it forms. This "interfacial free energy" component of surface modification reactions that controls the three-dimensional structure of the product has no counterpart in solution chemistry. Other unique considerations are the two-dimensional uniformity of reaction products and the uniformity of reaction as a function of depth into the solid.

The surface functionalization of unreactive polymers has advanced significantly in recent years. Fluoropolymers have been modified to incorporate discrete reactive functional groups (e.g., alcohols, amines, and carboxylic acids) into the outer few tens of angstroms of the solid samples. Polyethylene and polypropylene have been modified by selective oxidation, sulfonation, hydrogen abstraction/radical addition reactions, and "entrapment functionalization," in which functional groups are introduced to the surface by blending low-molecular-weight functionalized oligomers with the "host" polymer. The functional group chemistry of these modified polymers has been studied, and the control of surface properties, particularly wettability, with functional group manipulation has been assessed.

Reactive polymers that have been functionalized and the products well-characterized include poly(ether ether ketone) (the carbonyl chemistry was developed), polyimide (hydrolysis renders carboxylic acids), poly(phenylene terephthalamide) (metalation of the amide and subsequent alkylation), and polysulfone (metalation and carbonylation).

That the surface-modified materials are kinetic and not thermodynamic products is now recognized, and the thermal reconstructions of several functionalized polyolefin and fluoropolymer surfaces have been reported. The reconstructions typically involve movement of polar functional groups from the surface to greater depths and are driven by a decrease in surface energy. The functional groups do not migrate far and in some cases can be forced to return to the surface by contact with a polar medium. This mobility can effect desired surface properties and likely plays an important role in further chemistry of these surfaces.

Certainly significant advances have been made in this field in the past decade, but with respect to the highly evolved disciplines of organic polymer synthesis and synthetic organic chemistry, the field is in its infancy in terms of elegance and versatility. Few modified polymer surface systems have been studied with any degree of detail, and there is not currently a model system that can be used for comparisons. Key questions concern the applicability of the tenets of solution organic chemistry to surfaces, the utility of functionalized surfaces, correlations between surface structure and properties, and the new types of materials that can be prepared with surface synthetic techniques.

Biocatalysis in Polymer Synthesis

Organisms carry out an astonishing array of complex chemical transformations via coupled enzymatic reactions. The fact that enzymes operate selectively and under mild conditions of temperature, pressure, and solvent has generated justifiable interest in the use of enzymes in polymer synthesis. Work to date has addressed three issues: (1) in vitro use of isolated enzymes to catalyze polycondensations, (2) use of organisms (either wild type or genetically altered) to produce monomers that are subsequently converted to polymers by conventional methods, and (3) use of organisms to produce polymers directly.

The in vitro use of enzymes to catalyze polycondensations has been shown to offer several advantages. Enantioselective polymerizations have been reported, and polymerizations of monomers containing reactive functional groups (e.g., epoxy diesters) have been accomplished without destruction of the reactive functionality. Limitations of the method arise in part from the fact that many of the polymerizations of interest are best run in solvents for which the enzyme is poorly suited (i.e., in nonaqueous systems). As a result of this—and perhaps other—factors, molecular weights of the polymers prepared in this way are modest—generally $M_n < 15{,}000$. Recent developments in nonaqueous enzymology, including the use of protein engineering to improve enzymatic activity in organic solvents, offer promise for a solution to this problem.

Several intriguing reports have discussed the use of microorganisms to produce polymer intermediates. For example, 4,4′-dihydroxybiphenyl, an intermediate in the manufacture of engineering thermoplastics, has been produced in yields of greater than 95 percent in a one-step fermentation process. Microbial syntheses of long-chain dicarboxylic acids (useful in the production of polyamides and polyesters), and of 5,6-dihydroxy-1,3-cyclohexadiene (an intermediate in the preparation of polyparaphenylene) have also been reported. The opportunities for further development of this approach appear to be significant. Environmental considerations here are complex but will almost surely play a key role in defining the impact of biocatalysis in this area.

The direct production of polymers in bacteria and in higher plants is also under active investigation. Commercial development of biodegradable materials based on plant-derived polysaccharides has begun, and fermentation is being used to produce biodegradable polyesters for packaging applications. Recent reports of production of similar materials in higher plants suggest that agricultural routes to these polymers may ultimately prove viable.

A recent development has been the use of genetic engineering to produce polymeric materials. In some cases, this has taken the form of cloning and expression of the genes for natural structural proteins (e.g., spider silk); in others, amino acid copolymers have been designed de novo, encoded into artificial genes, and produced in bacterial hosts via fermentation. This approach offers significant advantages, in that it leads to uniform chain populations of controlled chain length, sequence, and stereochemistry, all the important structural components in polymer synthesis. In addition, the factors that control the secondary structure of proteins can be used to impose three-dimensional structure on the synthetic polymers made by this technique. This approach offers unique possibilities for tailoring the structure of synthetic polymers. The preparation of polymers with useful biological properties is particularly straightforward by this method, and there are promising indications that the scope of the method will be extended beyond the 20 naturally occurring amino acids. This area of research is an excellent point for the interface of the areas of biochemistry and traditional polymer chemistry.

Development of New Polymerization Methods

Even given the synthetic advances described in the preceding sections, there remains an urgent need for the development of new synthetic routes to polymeric materials. To illustrate, one need only consider the striking contrast between the synthetic methods used in the laboratory and in industry on the one hand and in nature on the other. In each case, the process begins with simple feedstocks, that is, with simple carbon compounds derived either from petroleum or from biomass. In industry, such compounds are isolated and then subjected to a series of separate chemical transformations leading to the monomer structure of interest. After rigorous purification to "polymerization grade" material, the monomer is converted in still another step, often at high temperature and pressure, to the polymer. Nature works differently. From mixed feedstocks, nature converts simple carbon compounds to complex polymers directly, without isolation of intermediates and without the assistance of harsh reaction conditions. Until industrial polymer chemistry works with the same degree of process integration and efficiency, there will be room for advances in synthetic methodology.

This analogy also serves to highlight the special value of synthetic methods that can make direct use of simple, inexpensive feedstocks for the synthesis of new polymers. An example of such an advance is the development of palladium-catalyzed copolymerizations of ethylene and carbon monoxide. As low-cost monomers, ethylene and carbon monoxide are difficult to beat. Early developments using radical polymerization techniques resulted in copolymers that were highly branched, with random monomer enchainment. Subsequent research demonstrated that palladium catalysts could be used to produce linear copolymers. These materials contained multiple ethylene insertion segments that resulted in a lack of photostability. Recent developments, derived from a better understanding of the basic chemistry of the catalysts, have resulted in a copolymer of ethylene and carbon monoxide that is perfectly alternating. The regular structure provides a crystalline material with excellent mechanical properties and enhanced photostability. To facilitate processing, the commercial product is a terpolymer of ethylene, carbon monoxide, and a few percent of propylene. This material is projected to capture some of the market that has been held by traditional polymers. The processes used to manufacture this polymer consume 1 gram (g) of palladium for every 10^6 g of polymer, resulting in a catalyst cost of $0.50 per ton. The commercial catalyst produces 3×10^4 g of polymer per gram of palladium per hour. A second example is provided by palladium-catalyzed carboalkoxylation and amidation reactions, which have been used to convert aryl halides to aromatic polymers in a single step, starting from carbon monoxide and the appropriate bisphenols or aryl diamines, respectively. New developments in catalysis and in organometallic chemistry will be critical to such discoveries and should be accorded high priority in exploratory research programs in polymer synthesis.

Exploring the Periodic Table: Inorganic Polymers

Although most current polymeric materials are based on the chemistry of carbon, the remainder of the periodic table is accessible through the synthesis of inorganic polymers and networks or hybrid organic-inorganic systems often composed of interpenetrating organic and inorganic polymers. Because the compositional and structural diversity of such nontraditional polymers is immense, only a few examples of synthesis strategies and challenges are presented here.

Inorganic and Organometallic Polymers

Inorganic and organometallic oligomers and polymers are macromolecules in which a metal or metalloid is part of the main-chain backbone or pendant to it. Familiar examples are those involving main-group elements: polyphosphazenes $[-RR'P{=}N-]_x$, polysiloxanes (silicones) $[-RR'SiO-]_x$, and polysilanes (polysilylenes) $[-RR'Si-]_x$ (Figure 4.10).

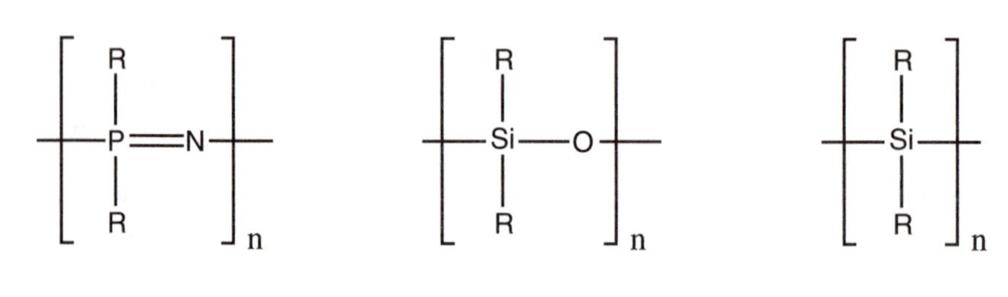

FIGURE 4.10 Three important classes of inorganic polymers; from left, the polyphosphazenes, the polysiloxanes, and the polysilanes.

The polyphosphazenes comprise a large and versatile class of inorganic macromolecules. Perhaps the most important feature of polyphosphazene chemistry is the method of synthesis that allows the R and R′ groups to be varied over a very wide range: organic, organometallic, or inorganic units, imparting diverse properties to the resulting polymers. In the macromolecular substitution method, hexachlorocyclotriphosphazene is heated to induce polymerization. Solutions of poly(dichlorophosphazene) in organic solvent are then reacted with nucleophiles such as alkali metal alkoxides, amines, or organometallics such as Grignard reagents to yield the derivative polymers plus the chlorides as by-products. Inherent to this synthesis procedure is the possibility that two or more different pendant groups can be introduced either simultaneously or sequentially. Organic or organometallic side groups can also be introduced at the cyclic trimer level, followed by ring-opening polymerization of the cyclic polymer to the high polymer. Because the substitutive method does not allow sufficient control over the stereoregularity of mixed-substituent polymers, microcrystallinity of polyphosphazenes occurs only when one type of side group is present.

Polysilanes and, analogously, polygermanes are usually made by dehalogenation of diorganodichlorosilanes with sodium metal in an inert diluent. Because cyclic oligomers are the exclusive products at equilibrium, high polymers are formed only when the reaction is kinetically controlled. At present, the reaction mechanisms are not well understood but probably involve the ion pair $[RR'SiCl]^- Na^+$, which in the rate-determining step reacts with dichlorosilane, adding one silicon unit and producing a chlorine-ended chain. Under ultrasound irradiation, at low temperature, high molecular weights and monomodal molecular weight distributions are obtained. Alternatively, dehydrogenation could be brought about catalytically by transition metals. Once formed, polysilanes may be chemically modified, or protected functional groups resistant to sodium metal may be used in the standard synthesis to introduce further compositional and structural diversity. A few copolymers have been synthesized. Silylene-olefin block copolymers can be made by the addition of styrene to anionically terminated polysilanes, followed by anionic polymerization of the olefin. Random polysilane-polygermane copolymers have been synthesized simply by co-condensation of dichlorosilanes and dichlorogermanes with elimination of sodium chloride.

The familiar polysiloxanes were first synthesized by hydrolysis of dichlorosilanes R_2SiCl_2, but this process has been largely replaced by ring-opening polymerization of cyclic oligomers, such as hexamethylcyclotrisiloxane, most often using anionic initiators. Initiation and propagation involve nucleophilic attack on the monomer, opening the ring, followed by chain extension. Polymerization of nonsymmetrical cyclic siloxanes gives polymers [–SiRR′O–] that are analogous to vinyl polymers. In principle, it should be possible to prepare them in stereoregular forms (isotactic and syndiotactic). This has not yet been accomplished.

There are of course many examples of less common inorganic polymers. Polymeric sulfur and selenium are synthesized by free radical polymerization of cyclic molecules. Stannoxane polymers with [–Sn–O–R′–] backbones are prepared by interfacial polymerization between diorganotin dihalides and difunctional organic molecules, while novel "drum" and "ladder" tin polymers are synthesized by the reaction of stannoic acid with a carboxylic acid. Addition or condensation polymerization can be used to prepare metal coordination polymers in which coordination occurs between an organic group in the polymer and a metal atom.

Network-forming Inorganic Polymers

Network-forming polymers incorporate branch points in the polymer backbone or on pendant groups. So-called "sol-gel" processing utilizing simple salts or metal organic precursors such as alkoxides $[M(OR)_n]$ has made much of the periodic table accessible to the synthetic inorganic chemist, although the distinction between polymer and ceramic now has become blurred. Sol-gel synthesis involves nucleophilic substitution reactions often represented as

$$M(OR)_n + mXOH \rightarrow [M(OR)_{n-m}\,(OX)_m] + mROH,$$

where X can be H (hydrolysis), M (condensation), or L (complexation). The chemical reactivity of a metal alkoxide toward hydrolysis and condensation increases with the positive charge of the metal and its ability to increase its coordination number. In general, the reactivity increases when going down a column of the periodic table ($Ti < Zr < Ce$). Because the preferred coordination numbers of most metals are much greater than two, the formation of polymeric networks (rather than precipitates) often requires complexation of the metal with slowly hydrolyzing multidentate ligands such as carboxylic acids, β-diketonates, or alcohol amines to reduce the effective functionality. Polymer size and structure are controlled by the L/M and H_2O/M ratios. Heterometallic alkoxides or oxoalkoxides are useful precursors to multicomponent polymers analogous to random copolymers and block copolymers. Recently, attempts have been made to synthesize well-defined oligomeric oxoalkoxides such as $Ti_{16}O_{16}(OEt)_{32}$ for use as "molecular building blocks" to assemble networks with intermediate range order.

Hybrid Organic-Inorganic Polymers and Networks

Structural features of traditional organic polymers are married with those of nontraditional inorganic polymers in hybrid materials. Polycarbosilanes [–SiRR′–CH_2–], polysilazanes [–SiRR′–NR″–], and polyborazines [–BR–NR′–], precursors to silicon carbide, silicon nitride, and boron nitride, also fall in this category. A common synthetic route to the formation of hybrid materials is the hydrolysis of organoalkoxysilanes $R'_xSi(OR)_{4-x}$ in a sol-gel process. If R′ is a nonreactive group such as an alkyl, it will serve to modify the inorganic network. If R′ is itself polymerizable (e.g., epoxy or vinyl), interpenetrating organic and inorganic networks can form. Depending on the choice of catalyst and the uniformity of the hydrolysis process, the reaction can be designed so that the organic and inorganic networks form simultaneously or sequentially. Other interesting hybrid materials result from the dispersion of an inorganic phase within an organic matrix or vice versa. For example, the swelling of polydimethylsiloxane with tetraethoxysilane followed by in situ hydrolysis yields a silicon-dioxide-filled composite. The addition of appropriate organic molecules or enzymes to sol-gel matrices results in optically active or bioactive materials.

Opportunities and Challenges

With most of the elements of the periodic table available, the opportunities for chemists to synthesize new inorganic polymers and networks with unique properties are clearly unlimited. However, the greater chemical and structural diversity represented by this class of materials compared to traditional organic polymers provides daunting synthetic challenges. In general, the ability to tailor chain (or network) architecture (e.g., topology, sequence, molecular weight, and stereochemistry) has not been widely demonstrated, especially for heterometallic systems. Future directions of research should focus not only on new materials but also on strategies to control the architectures of existing polymers and networks. Along these lines, some chemists are looking to nature, where there exist numerous examples of low-temperature routes to high-strength, high-toughness inorganic materials such as abalone shells and novel magnetic and nonlinear optical materials and processes such as magnetotactic growth of Fe_3O_4 in bacteria and cadmium sulfide particles formed by yeast. Biological systems rely on the organic matrix to control morphology and crystal nucleation sites, on interactive proteins to modulate crystal growth, and on ion transport systems (ion pumps) to provide localized supersaturated solutions. Other strategies that might provide more control over polymerization include nonhydrolytic processes, small molecule elimination reactions, polymerization within micelles and vesicles, and sonochemical syntheses.

Reactive Processing

The production of new materials depends not only on successful molecular design and construction, but also on effective control of processing steps. In many applications, coupled synthesis and processing schemes provide advantages.

Reaction injection molding (RIM) provides an example. RIM processes were introduced in the late 1960s and early 1970s and are widely used in the manufacture of polyurethane elastomers and foams. On a smaller scale, polyesters, epoxy resins, nylons, and dicyclopentadiene materials are produced in this way. The RIM technique involves rapid impingement of reactive monomers prior to injection into a mold where polymerization takes place, and the high rate of reaction between aromatic isocyanates and alcohols has made this chemistry particularly suitable for RIM technology. The challenge to the synthetic chemist is to develop new chemistries that occur at adequate rates in the absence of solvent and that are tolerant of the phase changes and temperature excursions that characterize the RIM process.

Reactive processing is also used to prepare a variety of toughened polymer blends. These processes use maleic anhydride-modified polyolefins or oxidized poly(phenylene oxide)s as reactive targets for covalent linkage with nylons. As in RIM, the chemistry must occur on the time scale of the processing step, and interesting challenges are posed by the interplay of chemical and physical factors (e.g., phase boundaries and thermal transitions) in determining the course of reaction and the properties of the product. New chemistries, both for preparation of reactive precursor polymers and for the interpolymer coupling step, are needed.

Supramolecular Chemistry

Polymer science and engineering have traditionally been concerned with the preparation and properties of chain molecules constructed via the formation of covalent bonds. Most often, such preparations involve reactions of organic monomers in isotropic phases such as solutions, suspensions, or emulsions. Over the last decade, there has emerged a broader appreciation of the role of noncovalent interactions in the synthesis of polymeric materials, and polymerizations in monolayers, bilayers, crystals, and liquid crystals have become a popular area of investigation. This work has been motivated by interest in (1) preorganization of the monomer as a strategy for controlling polymer structure and (2) use of polymerization to stabilize, or to alter the properties of, anisotropic phases.

Substantial successes have been achieved in these areas, but many aspects of polymerization in organized media remain poorly understood. It has been suggested that confinement of functional groups to a lattice or in two-dimensional arrays may either enhance or reduce reactivity, but the evidence for such effects is, with few exceptions, anecdotal. As materials science seeks control of structure and properties on shorter and shorter length scales, polymerization in anisotropic phases is sure to take on increasing importance.

A related area of investigation concerns polymeric structures held together by non-covalent forces, for example, by hydrogen bonds. Although they are significantly weaker than covalent bonds, hydrogen bonds exhibit directional character and are therefore useful in controlling both the size and the shape of molecular aggregates of dimensions comparable to those of polymer chains. Penetrating studies of molecular recognition processes in organic chemistry are providing new strategies for the assembly of large-scale structures, and it is clear that a broad view of polymer synthesis, a view that embraces both covalent and non-covalent bond-forming steps, should be encouraged.

Conclusions

- The last decade has witnessed advances in the control of macromolecular architecture. Nevertheless, important challenges remain, particularly with respect to the control of chain sequence. New synthetic approaches to controlled architecture, including the use of biological methods, should be vigorously pursued.
- Polymer synthesis, both on the laboratory scale and in manufacturing, is still largely a series of stepwise conversions, from feedstocks, to monomers, to polymers, to products. The development of more highly coupled, integrated syntheses, in which advanced materials are derived efficiently from simple feedstocks, will depend in large part on research on new catalysts and new catalytic reactions.
- The demands of new technologies will continue to place a premium on the synthesis of high-performance polymers, that is, on polymers of unprecedented thermal or mechanical properties. Continued efforts to design, prepare, and evaluate such materials must be accorded high priority.
- Polymer science in general—and polymer synthesis in particular—must embrace a broader view of the field, in which both covalent and non-covalent interactions are exploited to maximum advantage. This view will create lively exchanges of information and ideas between practitioners of polymer science and workers in related areas of organic and inorganic chemistry, biology, physics, engineering, and materials science. Work at these disciplinary interfaces should be supported and pursued.
- Environmental issues in polymer synthesis, as in other areas of polymer science, will continue to grow. Environmentally sound synthetic methods, strat-

egies for biomass conversion, and design for disposal all offer the synthetic chemist intriguing challenges as well as opportunities to contribute solutions to some of the nation's most urgent problems.

POLYMER CHARACTERIZATION

Significant advances in characterization techniques have paralleled the development of new polymeric materials. Identification of the unique properties and structure is necessary for further technological development and applications of the material. Sophisticated techniques now available to characterize various properties of polymers have dramatically improved our understanding of the mechanisms and principles governing polymer function. Using many techniques now developed for polymer characterization, individual molecules as well as molecular assemblies (i.e., melts and solids) and polymer surfaces and interfaces may be studied.

Rather than discuss all the techniques that have been introduced, many of which can be applied to characterize polymers in a wide variety of states, this section concentrates on recent breakthroughs and potential breakthroughs, where new capabilities should have major impact. A sampling of recent breakthroughs is given in Table 4.1, and potential breakthroughs are listed in Table 4.2. There are five main areas of application:

- *Molecular characterization of the architecture of isolated polymer chains;*
- *Characterization of solutions, melts, and elastomers,* especially the dynamics of polymer chains;
- *Characterization of polymer solid-state structure and properties*, wherein new microscopies can offer major advances;
- *Characterization of polymer surfaces and interfaces,* especially with new depth profiling techniques; and
- *Characterization of biopolymers* to provide information for development of biotechnology.

Molecular Characterization

Molecular characterization plays a critical role in understanding polymeric materials. New polymeric materials must be characterized to understand how their structure determines their properties; such information illuminates fundamental scientific questions and enables potential commercial applications.

Challenges are presented by synthetic homopolymers because they are generally polydisperse, not only in terms of molecular weight, but also in branching, tacticity, and microstructure. Copolymers may have all these polydispersities as well as additional ones in composition and sequence. Because physical proper-

TABLE 4.1 Recent Breakthroughs in Characterization of Polymers

1. Molecular Characterization
   - Size exclusion chromatography, combined with light scattering and viscosity detectors
   - Multidimensional nuclear magnetic resonance methods for determining chain architecture
   - Millisecond time-resolved ultraviolet/visible, Fourier transform infrared, and Raman spectroscopies
2. Solutions, Melts, and Elastomers
   - Many new techniques for measuring diffusion
   - Simultaneous rheological and optical/X-ray/neutron measurements
   - New nuclear magnetic resonance and optical techniques for measuring local polymer dynamics and tertiary structure of biopolymers
   - Neutron spin-echo techniques for measuring intermediate-scale polymer motion
   - Neutron-scattering methods for determining the thermodynamics of polymer blends
3. Solid-State Structure and Properties
   - Environmental scanning electron microscopy
   - Near-field optical microscopy
   - Simultaneous X-ray and calorimetry measurements
   - Solid-state nuclear magnetic resonance techniques
   - Molecular imaging with transmission electron microscopy
   - Confocal optical microscopy
   - Transmission electron microscope image-processing techniques
   - Techniques using synchrotron radiation sources for solution of the phase problem in solving the structure of large biomolecules
4. Surfaces and Interfaces
   - Many new deuterium depth-profiling techniques with complementary depth resolutions
   - Surface forces apparatus to characterize the forces between adsorbed polymer layers
   - Atomic force microscopy of surface topology
   - Neutron reflectometry of surfaces and interfaces
5. Biopolymers
   - Improved separation and purification methods and capillary zone electrophoresis for separating small amounts of biopolymers
   - Polymerase chain reaction technique for amplifying very small quantities of DNA
   - Multidimensional nuclear magnetic resonance methods for obtaining high-resolution structures in solution
   - Mass spectroscopy that quickly sequences biomolecules
   - Computer database technology for comparing informational macromolecules

ties of polymers can depend not only on the averages, but also on the polydispersities of the attributes just mentioned, it is necessary to characterize all of these. Methods for determining average molecular weights of soluble polymers and some of the other average properties are well known, but characterization methods for the distribution of these properties, especially for more than one of these properties at a time, are in short supply.

TABLE 4.2 Potential Breakthroughs in Characterization of Polymers

1. Molecular Characterization
   - Ways to find molecular weight distributions for insoluble polymers
   - Better ways to determine branch content and sequence information
   - Better ways of measuring how branch content and monomer composition are distributed over range of molecular weights
2. Solutions, Melts, and Elastomers
   - Better ways to measure the distribution of molecular lengths between cross-links
   - More intense sources of long-wavelength neutrons to make possible time-resolved experiments
   - Better techniques for the preparation of isotopically labeled biomaterials
3. Solid-State Structure and Properties
   - Energy-filtered soft X-ray microscopy
   - Electron tomography
   - New techniques for reducing electron radiation damage
   - Chemical-imaging transmission electron microscopy
   - Ways to measure plastic deformation and fracture properties of small samples
   - Methods to probe mechanism of shear yielding in glassy polymers
   - Methods to characterize the microscopic deformation of rubber-modified polymers at impact strain rates
   - New techniques for the solution of the phase problem in biomolecules
   - Computational techniques for the solution of nonglobular virus structures such as the AIDS virus and complex structures such as those of molecular motors
4. Surfaces and Interfaces
   - Analysis techniques that can characterize curved interfaces
   - Interface analysis techniques with both good depth resolution and good lateral resolution
   - Ways to use atomic force microscopy to characterize local mechanical properties
5. Biopolymers
   - Scanning tunneling microscopy to read out biopolymer sequences
   - Rapid sequencing methods
   - Computer algorithms to predict biomolecular structures from sequences
   - Higher-resolution electrophoretic methods for separating large biopolymers
   - Time-resolved Laue diffraction methods for X-ray crystal structure determination
   - Hydrogen exchange nuclear magnetic resonance for local motions and internal structure of biopolymers

The number and average molecular weight of soluble polymers can be obtained by numerous methods, including colligative property measurements, scattering, and ultracentrifugation. Viscosity measurements, which are sensitive to molecular weight, size, and polydispersity, are also useful. Size exclusion chromatography (SEC), which is sensitive to molecular size rather than weight, now may be equipped with light scattering and viscosity detectors so that absolute molecular weights can be obtained directly. Further research is needed, however, to improve the material used to pack the SEC columns. Better columns are needed both for the analysis of high-molecular-weight samples (>10^7) and for use at high temperatures. Methods for obtaining molecular weight distributions of insoluble polymers are, however, in their infancy; one example involves mass spectroscopy, which has recently been used for molecules with molecular weights as high as 100,000. Methods for studying higher-molecular-weight molecules are needed.

There are few methods for studying more than one polydispersity at a time. Among the few existing methods is orthogonal chromatography, in which the output of a SEC-yielding separation by molecular size is injected into a second SEC or a high-performance liquid chromatograph operated so as to separate by composition. Routine methods for providing such analysis need to be developed.

All instrumental methods are benefiting from the advances in electronics and data-handling techniques that have occurred over the last three decades, and this progress continues apace. The improved sensitivity has led to shorter sampling times, allowing study of kinetic properties, and higher spatial resolution, which should enable the development of new forms of microscopy and imaging.

Nuclear magnetic resonance (NMR) spectroscopy is still the premier tool for investigating the architecture of the polymer chain, that is, monomer sequence, stereochemistry and tacticity, monomer orientation, and branch content. The power of NMR arises because the signals for the different hydrogen, carbon, and other atoms along the main chain and side chains of polymers can be resolved, and the NMR parameters (chemical shifts, coupling constants, and relaxation times) are determined by the local environment. Recent advances in NMR are higher-field magnets leading to increased sensitivity and resolution and the development of new methods to study solid polymers. The introduction of multidimensional NMR (NMR experiments with more than one frequency dimension) has been a particularly important development. Prior to the development of such methods, it was not possible to study the interactions between many pairs of carbons or protons because of spectral overlap. These interactions can be visualized using two-dimensional NMR because the spectra are dispersed into a two-dimensional plane rather than along a single-dimensional axis. Similar multidimensional NMR techniques have been discovered that make it possible to investigate polymer chain and side group motions in much more detail than in the past. The use of solid-state NMR allows one to analyze bulk polymers and films, as well as insoluble materials. NMR can also be used to study the local organization in semicrystalline materials and the domain size in phase-separated materials.

While NMR techniques have made it possible to routinely determine the average tacticity, microstructure, and composition in polymers, NMR methods for studying the monomer sequence and branching still can yield only a partial description of these. It is currently much too difficult to obtain information on the distribution of these attributes in a sample by NMR or otherwise.

We are all familiar with the wonders that three-dimensional NMR imaging has wrought for medical diagnosis; such imaging has great potential for polymer science as well. At the 1- to 0.1-millimeter (mm) level already achieved in medical instruments, NMR imaging can potentially play a role in characterizing polymer processes such as injection molding, mixing, and extrusion. While at the moment three-dimensional NMR imaging even at the scale of an optical microscope (10 to 1 micrometer [μm]) seems far in the future because of sensitivity problems, one- and two-dimensional NMR microscopies may be imminent. Such microscopies, with their sensitivity to details of chemical bonding and isotopic composition, will find relevant applications in characterizing diffusion, reaction, and structures in polymer solutions, melts, and solids.

Other spectroscopic methods of importance to polymer researchers are ultraviolet/visible (UV/Vis), Fourier transform infrared (FTIR), and Raman spectroscopies. These spectroscopies, in common with NMR, are nondestructive; no changes in the chemical bonding, conformational order, crystalline structure, or morphology of the polymer are caused during the interrogation of the sample.

UV/Vis spectroscopy provides information on the electronic states of the polymer backbone in both crystalline and amorphous polymer phases. It has played a major role in unraveling the mechanism of conduction and nonlinear optical response in π-conjugated polymers while providing new insights into thermo- and piezochromic behavior in σ-conjugated systems. Recent advances allow UV/Vis spectra to be obtained in less than 100 milliseconds (ms). Future advancements will focus on high-sensitivity diode array detectors and even faster data acquisition times providing UV/Vis spectra "on the fly" so as to follow reaction kinetics and phase transitions.

Probably the most revolutionary optical spectroscopic advance in the polymer area in the last decade has been the ability of Raman spectroscopy to provide important conformational and morphological information about polymers. Not only have the detectors used in conventional Raman spectrometers been improved significantly from the former single-element photomultiplier to a multipixel array charge-coupled detector (CCD), but their range of sensitivity has also extended to the near infrared with the introduction of an entirely new set of solid-state lasers. This development has provided new pathways for the investigation of polymers that contain intrinsic chromophores as part of their structure, and following the introduction of Fourier transform spectroscopy, structural polymers containing fluorescent impurities can still be investigated using Raman spectroscopy without the fluorescence overwhelming the much weaker inelastically scattered Raman signal. At an incident wavelength of 1,064 nanometers (nm), no fluorescence can be excited because the incident photon energy is too low, and hence good-quality Raman data can be routinely collected. With significant improvement in detectors on the horizon, Fourier transform-Raman and conventional Raman (using visible excitation) will mount a formidable challenge to FTIR spectroscopy in the area of sensitivity.

Characterization of Solutions, Melts, and Elastomers

The past decade has brought major advances in methods for characterizing solutions and melts, particularly techniques for investigating "molecular rheology," the relation between the molecular structure of a polymer and its flow characteristics. This area is particularly important because the main polymer-processing methods, including injection molding, fiber spinning, and extrusion, involve flow of melts or solutions. Many new techniques for studying diffusion have been invented, and diffusion measurements have become a major probe of molecular rheology. Some of these techniques, such as fluorescence recovery after photobleaching or forced Rayleigh scattering, involve polymers labeled with fluorescent molecules that can be bleached or chemically altered with an optical interference pattern; analysis of the time dependence of the disappearance of the bleached pattern yields a diffusion coefficient. Other techniques rely on labeling of the diffusing polymers with isotopes, usually deuterium. The concentration profile of the deuterium-labeled polymer diffusing in a protonated polymer melt can be determined directly by infrared microdensitometry or by any of a number of depth-profiling methods, as discussed below.

Rheological characterization capabilities have also expanded greatly. Particularly promising are rheometers that allow optical techniques (e.g., infrared dichroism and light scattering) that probe the deformation of polymer molecules and phases in melts and solutions to be carried out under well-defined stress and strain rate conditions. Recent studies of light scattering from sheared solutions and blends, for example, have demonstrated the effect of shear on fluctuations and phase stability. These can be expected to be critical in defining processing conditions to produce desired phase structures. Other developments of note have included the use of synchrotron X-ray beams to probe the rheological state of liquid crystalline melts and solutions. The high intensity of such beams makes possible characterization not only of the steady-state structure of the system under flow but also of transient structures. Major advances in this area await the development of fast area detectors for X-rays that can be read and recorded on a time scale of seconds or even milliseconds.

Neutron scattering has become an indispensable tool in the characterization of polymer melts and elastomeric networks. Neutrons of wavelengths from 0.2 to 2.5 nm available from modern neutron sources permit studies of the dimensions of single polymer molecules as well as the chain conformation over much shorter length scales.

Neutron scattering can take great advantage of deuterium-labeled polymers in which deuterium has been substituted for hydrogen. Because of the great difference in neutron scattering length between these isotopes, appreciable contrast between hydrogen- and deuterium-containing molecules may be achieved. While little chemical or thermodynamic perturbation is produced by the replacement of hydrogen by deuterium, the effect is important in some situations. The contrast permits the measurement of the geometry of single molecules in concentrated solutions and melts, of the topology of molecules folded within crystals or interconnecting them, and of the connectivity of microdomains in block copolymers. Selective substitution in parts of molecules allows for the determination of their arrangement in superstructures. Portions of structural units, such as those of shell-core structures and micelles, may be labeled as an aid to their structural determination. The extension of single molecules in sheared melts and stretched networks can be measured, permitting critical tests of deformation theories. However, better methods are needed to characterize the distribution of molecular lengths between cross-links in networks produced under polymer-processing conditions typical of industrial practice.

Through use of methods such as neutron spin-echo spectrometry (Mesei, 1980), experiments have provided the most thorough test to date of whether the details of the reptation model are correct (Richter et al., 1990; Butera et al., 1991). Quasi-elastic neutron scattering has extended the range of dynamic studies achievable by light scattering to smaller dimensions, utilizing the advantages of isotopic labeling. These advantages are somewhat offset by the low neutron fluxes available, leading to long data acquisition times.

The neutron scattering from polymer mixtures depends on fluctuations from uniformity. Such concentration fluctuations require work against osmotic forces, and so their size relates to the osmotic compressibility. Thus, neutron-scattering measurements can provide thermodynamic information. While such techniques have long been applied to the molecular characterization of dilute polymer solutions, recent developments in neutron scattering using deuterium-labeled molecules have permitted their extension to concentrated solutions, amorphous blends, and melts. These have provided values of interaction parameters (Flory χ values) in such concentrated systems. The excess scattering occurring as phase separation is approached serves to characterize critical phenomena and has proved useful for studying spinodal decomposition of polymer mixtures.

Scattering techniques have grown in use because of developments of intense radiation sources (lasers, synchrotrons), of efficient area detectors (optical multichannel analyzers, CCDs), and the employment of computers for rapid data analysis. Faster X-ray detectors are needed for efficient use of the high fluxes available with synchrotrons. Currently, available neutron fluxes are too low to permit dynamic studies, so that the prospect of a high-flux advanced neutron source is a welcome possibility. The case for an advanced neutron source is reinforced by the emergence of neutron reflectometry as a primary tool for characterizing polymer surfaces and interfaces (see below), because our aging reactor neutron sources are nearing the end of the time when they can be operated safely to give even-current fluxes of neutrons. Synchrotron X-ray and neutron-scattering techniques require experimentation at a central facility, and therefore the development of user-oriented centers to permit measurements by users other than the experts in these fields is highly desirable so as to encourage more general use of these important techniques.

Characterization of Polymer Solid-State Structure and Properties

Structural characterization of the polymer solid state has advanced rapidly, driven partly by the need to understand the complicated structures present in semicrystalline polymers and partly by the desire to characterize the morphology of multiphase polymer blends. Traditional thermal analysis techniques, such as differential scanning calorimetry, have been combined with powerful synchrotron X-ray sources to allow simultaneous structure and calorimetric measurements in a time-resolved fashion. The structural transformation giving rise to a thermal effect thus can be unambiguously determined.

Many new microscopy techniques, developed in the last decade, are just beginning to be exploited by polymer scientists. For example, it is difficult to observe thick specimens owing to the limited depth of field of conventional optical microscopes. By focusing a laser beam on a particular point on the sample and scanning, an "in focus" three-dimensional image can be constructed and an image of a section at a given depth can be generated. Resolution can be enhanced by using fluorescent dyes that require a two-photon process for emission.

Near-field scanning optical microscopy (NSOM) is another new technique that has found important applications in biology and that could be very useful for high-resolution optical microscopy of polymers. An evanescent optical wave field from the tip of an optical fiber, as small as 12 nm in diameter and much smaller than the wavelength of the light, is coupled to the surface of the sample, exciting fluorescence or scattering. As the tip is scanned over the surface of the sample, the image thus produced has a resolution much better than that of conventional optical microscopes, but without the problems (such as radiation damage and vacuum) associated with electron microscopies.

New environmental scanning electron microscopes (SEMs), which can operate in a partial pressure of water vapor, can eliminate the necessity for coating samples with metal to prevent charging artifacts, thus achieving superior resolution. Solvent as well as water effects on the mechanical properties can be studied in situ.

For the ultimate in resolution, the transmission electron microscope (TEM) is necessary, but until recently TEM images of polymers were restricted to relatively low magnification because of radiation damage effects. Recently, however, imaging of molecular structures in radiation-resistant polymers has been demonstrated by using TEM image-processing methods. Biologists have made use of electron tomography (the electron equivalent of the X-ray CAT scan) to reconstruct three-dimensional images of structures within the cell. A number of polymer problems would benefit from a similar approach.

The ability to generate high-resolution chemical maps of the polymer structure in blends would be a breakthrough. Current scanning transmission electron microscopic methods, which employ energy-dispersive X-ray analysis or electron energy loss spectroscopy, are not suitable for most polymers. New chemical imaging electron microscopes, which permit recording of images created with electrons with a well-defined energy loss, may revolutionize this field. By digitally subtracting two images, one just below and one just above the absorption edge of a particular element, one can generate a chemical map of that element, while minimizing radiation damage.

A related technique for chemical mapping is soft X-ray microscopy. By tuning the X-ray energy to a strong absorption characteristic of a particular moiety, say, the π-plasmon associated with a phenyl group on one of the components of the blend, and scanning the focused X-ray beam, one can map the distribution of that moiety over the sample. Tunable soft X-ray sources are available at synchrotron radiation facilities, and the necessary focusing X-ray optics are now becoming available; already, spatial resolutions of approximately 50 nm have been demonstrated.

For most of the applications of solid polymers, mechanical properties are of primary importance. While there have been advances in characterizing and understanding these properties, there are many areas where improvements can be made. Techniques to characterize the mechanical properties of small samples of polymers are needed. Most conventional mechanical testing requires very large samples, for example, compact tension samples for fracture toughness measurements, tensile "dogbone" bars, and so on. To evaluate a new polymer or polymer blend, one would like to use 10-gram (g) batches synthesized by a polymer chemist rather than scaling up to produce 10 kilograms (kg) in a pilot plant. It is now possible to test polymer films that are as thin as 1 μm; the principal difficulty is that thickness has well-known and severe effects on the mechanics of the plastic deformation of polymers. Nevertheless, with experience it is possible to extrapolate from measurements on films to bulk samples. Thin film testing has the natural advantage that samples can be examined by both optical microscopy and transmission electron microscopy. Another area where improvements can be made rapidly is in fracture testing. The fracture behavior of interfaces and thin polymer layers embedded between two tough, transparent polymer slabs may be measured by using nothing more than a razor blade and an optical microscope. Further development to allow opaque polymers to be tested would be of great benefit.

While there has been considerable progress in the past several decades in our fundamental understanding of the deformation and fracture of solid polymers, particularly in the area of crazing and fracture, much more improvement is possible. The microscopic mechanisms of shear deformation are little known, yet these mechanisms are responsible for the toughness of the toughest polymers, such as polycarbonate. The addition of rubber particles is the usual way of producing tough polymeric materials, yet the microscopic fundamentals of rubber toughening are not yet fully understood. Experimental techniques to probe the microscopic nature of the deformation and fracture of such systems, preferably in a time-resolved fashion at typical impact strain rates, are needed.

Characterization of Polymer Surfaces and Interfaces

Surfaces and interfaces present challenges that are distinctly different from those presented by bulk, three-dimensional polymeric materials. The surface or interface introduces new characteristics unique to its quasi-two-dimensional character. They require special techniques that sample only the layer of interest near the top surface or at the buried interface.

New depth-profiling techniques, which measure some attribute of the polymer as a function of depth from the outer surface, are advancing this field rapidly. While some well-established techniques, such as angle-resolved X-ray photoelectron spectrometry (XPS), can sense the local chemistry near the surface, others, such as the ion spectrometries, forward recoil spectrometry (FRES), nuclear reaction analysis (NRA), and secondary ion mass spectrometry (SIMS), as well as neutron reflectivity (NR) and attenuated total reflection FTIR (Fourier transform infrared) spectroscopy, can find the depth profile of deuterium-labeled polymers, with depth resolutions usually well below 100 nm. These depth-profiling techniques have been used recently to characterize short-range diffusion of polymers across interfaces, the enrichment at the surface of one component in a miscible polymer blend, the depletion of polymers near a solution-solid interface, and the segregation of block copolymer surfactants to the interface between two immiscible polymers, to name only a few applications. It should be realized that the techniques have complementary capabilities, with some, such as NR, providing the best resolution of sharp features of the profile and others, such as FRES, enabling a better picture to be drawn of their long-range and integrated features.

While depth profiling is discovering aspects of the structure of polymer surfaces and interfaces never before accessible, it is also likely to have a major impact on our understanding of polymer fracture. Depth profiling the two surfaces produced by fracture has made it possible to determine the locus of fracture precisely with respect to particular deuterium-labeled polymers (a deuterium-labeled block of a diblock copolymer) and to derive great insight with regard to mechanism of fracture from this information.

Such depth-profiling measurements, however, usually require flat samples, which are stratified in depth; making such measurements on samples with geometries such as cylinders (fibers) and spheres (particles) is normally not possible. In addition, all of these techniques have much more limited resolution laterally, that is, in directions in the plane of the interface. These lateral resolutions range from centimeters to, at best, micrometers. There is a need for interface analysis techniques that have both good depth and good lateral resolution and that can analyze curved interfaces.

Because the different depth-profiling techniques have complementary capabilities and no one technique is usually adequate to provide the level of resolution, sensitivity, and quantitation desired, it is important that researchers in this field have access to a wide variety of depth-profiling instrumentation. Facilities for SIMS and ion beam analysis, while common in semiconductor research, are not widely available to researchers interested in polymers. Some effort, perhaps on a regional basis and involving both universities and industry, should be made to ensure that this instrumentation is accessible to polymer scientists.

The characterization of interfacial properties has also advanced dramatically in the past decade. The surface forces apparatus (SFA) has become a standard method for measuring the (normal) forces between layers of polymer adsorbed in solution on solid substrates. Such measurements are important for understanding the steric stabilization of colloidal suspensions by polymers as well as for determining the thermodynamics of polymer adsorption from solution. For example, such measurements have led to an understanding of the stretching of polymer chains normal to interfaces between a solid and the solution as a function of their areal density. While the original SFA required the use of cleaved mica surfaces, recent workers have lifted this restriction, and a much wider variety of solid surfaces is now available. Shear forces between surfaces, as well as normal forces, can now be measured by some SFA devices, enabling the study of the fundamentals of polymer effects on lubrication. The SFA remains an apparatus that demands considerable skill and experience on the part of the experimenter, however; efforts to simplify the apparatus by improving the drive and mechanisms for detecting distance are under way and should be encouraged.

The SFA has also been used in the last few years to measure interfacial energies and adhesion; by measuring the area of contact between the polymer-coated mica cylinders as a function of the applied normal force, one can determine the reversible work of adhesion. An elegant variant of this method has been developed recently using a elastomer sphere pressed into contact with a elastomer flat. The surface of the elastomer can be modified—for example, one can form a thin silicon dioxide layer on a siloxane elastomer by treatment with an oxygen plasma—and highly organized organic films can be self-assembled on such modified surfaces. The work of adhesion between a wide variety of surfaces can thus be measured with little more than an optical microscope and an analytical balance.

More conventional ways of measuring surface properties include the determination of contact angles of fluids on polymer surfaces and the measurement of the shapes of polymer melt and solution droplets, under conditions where these shapes are modified by the action of gravity or centrifugal force. Here the major difficulties are caused by the highly viscous nature of polymer melts, which makes it hard to know whether the equilibrium shape has been obtained. It is particularly difficult to measure the interfacial tension between phases in a polymer melt. One promising approach involves measuring the retraction and breakup into droplets of a fiber of one phase in a matrix of another, but more attention should be devoted to this problem because the interfacial tension helps determine the morphology of melt-processed polymer multiphase blends.

The atomic force microscope (AFM) has already had an impact on the measurement of the surface topology of polymers. It seems likely, however, that the AFM can be suitably modified to allow measurement of very local surface properties as well as mechanical properties of polymers. One example is the lateral force microscope (LFM), an AFM modified so it can measure lateral forces as well as normal forces. With the LFM, it is possible to reveal the morphology of a phase-separated blend of hydrocarbon elastomers simply by scanning a microtomed surface. Another, more speculative, possibility would be the chemically sensitive AFM. In such an instrument, the tip would be modified to expose an outer surface of hydrogen bonding groups. By operating such an AFM in a "tapping" mode, it might be possible to distinguish hydrogen bonding regions of the surface from nonhydrogen bonding ones. Such further developments of the AFM and related instruments should have a large payoff in polymer surface research.

Characterization of Biopolymers

Biopolymers require many techniques other than those used for synthetic polymers. To characterize a biopolymer, the first steps are to purify and produce the material in quantity and learn its molecular sequence. The main methods for sequencing and purification include chromatography, gel electrophoresis, centrifugation, and dialysis. They will continue to play important roles. There are also new methods that promise to revolutionize how we obtain, purify, and understand the message encoded in biopolymer sequences. Capillary zone electrophoresis separates materials with very high resolution and works with extremely small volumes of material (in some cases, even the volume of a single biological cell!). Mass spectrometry, particularly in conjunction with electrospray and matrix-assisted laser desorption methods, can determine the sequences of informational polymers very rapidly and can detect subtle aspects of biologically important sequences. At present, there are limitations on the sizes of molecules that can be studied in this way, but the maximum size is growing as the technology evolves. Polymerase chain reaction (PCR) is a method for amplifying very small quantities of DNA. Because most molecules in the cell are in limited supply, this technology now opens the possibility of fishing out even the rarest of molecules and producing the appropriate DNA, RNA, or protein sequence in sufficient quantity to study them. Major advances in identifying and characterizing new proteins and nucleic acids will result from new electrophoretic techniques with enhanced resolution and reproducibility. Having determined the sequence of a biopolymer, computer database search methods now make it possible to search through the enormous number of known sequences and structures to make educated guesses about the biopolymer's possible structure and function. Because sequencing and purification methods are developing rapidly, the sizes of the databases are growing exponentially. The sophistication of computer search methods is keeping pace with this information explosion.

Structure Determination

Once a sufficient quantity of a pure biopolymer is available, the next step is to learn its atomic structure. The main methods for determining biopolymer structures are X-ray crystallography, NMR spectroscopic methods, electron microscopy, and scanning tunneling microscopy. The highest-resolution structures have been obtained by X-ray crystallography. Because crystallography requires that the molecules be crystallized, which is a poorly understood process, obtaining new structures by crystallography can be slow. A further bottleneck in obtaining crystal structures is the so-called "phase problem," which results from the unavoidable loss of information in recording X-ray intensities rather than amplitudes and makes it difficult to reconstruct the molecular structure from the diffraction data. While this problem is most severe for the large molecules typical of biopolymers, new approaches and solutions continue to emerge. Multiple wavelength anomalous diffraction, a technique requiring a synchrotron source, provides one solution to the phase problem. A most exciting new technology is the Laue diffraction method for time-resolved X-ray crystallography. Preliminary successes indicate that this technique may allow us to watch the dynamic processes, step-by-step, in chemical reactions and physical changes in biomolecules.

Another major development has been the emergence of multidimensional NMR for obtaining high-resolution structures of biomolecules in solution. An advantage over crystallography is that the NMR experiment does not involve crystallization or the phase problem. However, NMR is currently limited to studying smaller molecules than X-ray crystallography, although this situation is rapidly improving. NMR imaging of supermolecular biopolymer structures would offer a tool with almost limitless possibilities. Major advances could result from the ability to label biomolecules cheaply. Carbon-13 NMR spectroscopy of proteins is extremely powerful. However, its use is limited owing to the present expense of preparing the required carbon-13-enriched samples. Scanning tunneling microscopy is another new technology with preliminary successes and much promise for biomolecular structures.

The Folding Problem

How do biopolymer structures arise from interatomic forces? A major goal in characterization is to explain biomolecular structure using fundamental chemical principles. The "folding problem" is to understand how the monomer sequence encodes the conformation of the biopolymer. The fact that molecular "chaperones" are needed to obtain the correct folded structure for some proteins complicates the issue even further. Major efforts are going into developing theoretical models and methods of computational chemistry and statistical mechanics to allow one to predict the folding. The two main problems are the need for better interatomic and intermolecular potential functions, particularly for modeling aqueous solutions, and faster conformational search strategies, which will also rely on new developments in massively parallel computer hardware and software, in order to find conformations of low free energy. Experimental advances have been occurring rapidly in several areas. Mutation experiments replace one or more monomers, often through genetic engineering, and explore the consequences of the small changes on structure or function of a biopolymer. Unnatural monomers are synthesized chemically to create mimics of biomolecules or monomer replacements. Hydrogen exchange NMR methods have recently begun to allow unprecedented access to information about the local motions and internal structure. Small-molecule models and high-resolution calorimetry are helping to understand the driving forces. Direct synthesis and redesign of proteins are now becoming feasible, for increased stability and new applications of proteins as catalysts, and in information and energy transduction and storage.

Conclusions

- National facilities are important for polymer characterization. These facilities, producing neutron and synchrotron radiation, are employed for diverse purposes, and continuing support and improvement must be based on the sum of the benefits and costs.
- Improved characterization methods for structures between the monomeric and macroscopic levels are needed, for solutions, surfaces, solid-state polymers, and insoluble polymeric materials. Methods and instrumentation need to be developed.
- Polymer characterization has prospered from the collaborative interaction of various disciplines with the common goal of developing characterization tools. While many techniques have been specially developed for polymers, many were imported into polymer science after being developed for other purposes. Given the extent to which polymer characterization has been advanced by techniques developed first in other fields, the education of polymer scientists and engineers must be broad enough that they will be able to capitalize on similar

outside developments in the future. At the same time, increased contact between polymer workers and those in outside fields will ensure that new techniques developed within polymers will be rapidly exploited in those outside fields where they are useful.

THEORY, MODELING, AND SIMULATION

Enormous opportunities and new needs for polymeric materials exist in the rapidly expanding and internationally competitive high-technology areas, such as the information industry, aerospace industry, and biotechnology. Realizing these opportunities and meeting the needs for novel materials, however, depend on a much deeper understanding of polymeric materials than has been necessary for developing commodity polymers in the traditional chemical industry.

Theory and computations can help provide this deeper understanding. They reduce large collections of experimental observations to working knowledge, rules, patterns, models, and general understanding. They explain experimental observations; they correlate data from different materials and phenomena; and, in general, they quantify and unify our knowledge. More importantly, theory provides new predictive capabilities to guide the development of new ideas and to direct experimental efforts in exploring new chemical structures, processes, and physical properties.

In favorable cases, theory and simulation can reduce the amount of experimental work required or even eliminate it entirely. For example, one of the classical theoretical problems is to predict "phase diagrams," diagrams that describe the resulting states of matter and the conditions when polymers and solvents are mixed together. This is a central problem in the design of new materials. Modern polymeric materials often involve mixtures of several different types of polymers, of different molecular weights, and in complex solvents at different temperatures and pressures. They involve many variables. Such variables determine the difference between achieving a successful material with appropriate optical, thermal, mechanical, and chemical stability properties or producing a useless mixture. To find optimal conditions could require exploring 5 or 10 variables in detail, which could require tens of person-years of experiments. Instead, theoretical or computational models often permit the exploration of many different variables quickly, and thus they reduce the amount of time to develop new materials by an enormous factor.

Another example of the use of theory and simulation is in the area of polymer processing. Many companies currently model (1) kinetics of polymerization reactions and (2) fabrication, combining phenomenological rheological and heat transfer characterization of the polymer, to assess a proposed design before cutting a mold or die. Extensive efforts have been devoted to modeling fabrication operations such as extrusion, mixing, and molding operations. Once a process configuration has been selected, computer-based models are generally able to predict temperature, pressure, and conversion profiles in time or space. More sophisticated versions are able to predict certain product qualities, such as molecular weight averages, copolymer composition, and branching frequencies. They can sometimes also evaluate questions of process safety. (Additional modeling information is given in Chapter 3 in the section "Polymer Processing.")

While even rudimentary theories are often important sources of guidance, the ultimate power of theoretical and computational models depends on reducing their assumptions and approximations by building deeper and more fundamental models.

Theory and computer modeling will play increasingly valuable roles in the development of new advanced materials. The strong interdependence of experiment and theory in polymer science is well illustrated by the Nobel Prizes in polymer theory awarded in 1974 to Paul J. Flory (in chemistry) and in 1991 to Pierre-Gilles de Gennes (in physics). In certain areas of polymer theory, there has been enormous progress in the 1970s and 1980s. Some topics have been so thoroughly explored that they are currently receiving little attention, while other areas have moved to the forefront of theoretical activity. Further shifts in the centers of interest will undoubtedly be propelled by advances in theoretical methods, experimental developments, societal needs, and the phenomenal increase in computer performance and algorithm development.

Theory and computation have developed into two separate disciplines, although they are often intertwined to varying degrees. Theory is associated with the construction of physical and mathematical models of the system, with attempts to solve basic equations describing the properties of the system, and with predictions for the outcomes of experiments, including those that have not yet been done. Computation represents an application of theory, often with the aid of large-scale computers, either to compute the desired properties for a system of interest or to simulate the behavior of the system at a level of detail unavailable experimentally. These simulations are designed for furthering our understanding to test current theories or to provide data necessary for developing new and improved theories.

Great advances in theory and in computational resources have resulted in the development of a scientific computation industry, which constructs large-scale computer codes that are increasingly being used by scientists and engineers in industry and universities, in addition to the home-grown codes, which are in a constant state of renewal and development. It must be stressed, however, that these computer packages are no better than their underlying theories and the input data employed. Thus, advances in theoretical methodologies, computer technology, and the acquisition and codification of relevant experimental data will play essential roles in improving the capabilities of these computer packages.

Theory and computation are applied on several levels, ranging from the microscopic level, involving a detailed description of the molecular constituents, to a macroscopic level, involving continuum mechanics or thermodynamics, to describe the bulk properties of matter. Intermediate between the microscopic and macroscopic descriptions lies the mesoscopic world characterized by lengths long on a microscopic scale but short on a macroscopic one. Each level generates its own conceptual, theoretical, and computational challenges, but perhaps the most significant challange lies in providing a bridge between the different levels. Thus, a goal of the complete molecular modeling of materials involves use of molecular theories to compute the property information necessary to describe the processing or performance behavior of the new bulk materials.

States of Matter

Polymer Solutions

Because many theories of polymers in concentrated solutions and bulk rely on single-chain concepts developed and tested in dilute solutions, the understanding of dilute polymer solutions has repercussions throughout polymer science. Hence, dilute solutions of flexible polymers have received much attention in the past 50 years, and as a result many aspects of the average conformation of an isolated polymer molecule are now well understood. Nevertheless, many problems remain, especially for solutions of stiff polymers and their hydrodynamic properties. As the solution concentration increases, these theoretical problems become even more complex, and these properties are less well understood.

Amorphous Polymers

Flexible polymer molecules in the undiluted amorphous state generally assume unperturbed random-coil configurations. This fundamental conclusion is derived from both theoretical and experimental studies on very flexible and nonpolar polymers. However, many technologically important polymers normally contain considerable numbers of polar atoms or groups and exhibit considerably reduced chain flexibility. Because sufficient chain stiffness and/or orientation-dependent polar interatomic interactions produce anisotropic ordered phases, the embryonic structure of such order is present in amorphous polymers, which are not very flexible or which contain polar groups, as indicated by recent computer simulations. Simulations and theoretical studies are needed, therefore, of the local correlations among such molecules in the amorphous state to guide and interpret suitable experimental studies. An amorphous polymer is a glass below its glass transition temperature, T_g. Such glasses are never in thermodynamic equilibrium, and consequently their properties depend on their thermal and mechanical histories. Although significant efforts have been made recently in both theoretical and simulation studies of the glass transition, and new computational advances have opened the way to simulation of the properties of the glassy state, much more effort is needed to explain the underlying cause of the glass transition, how it depends on the chemical structure of a polymer, and the specific characteristics of glasses that emerge from polymers. Above T_g, thermal energy produces long-range molecular motions whose rate increases rapidly with the temperature. At sufficiently high temperatures, a polymer is either an elastic fluid or an elastomer, depending on whether or not it is cross-linked into a three-dimensional network structure. When the elastomer chains are sufficiently long, its elastic behavior is explained on a single-chain molecular level. However, our understanding of the role played by trapped entanglements and the swelling behavior of networks due to solvent remains limited.

Crystalline Polymers

Crystallinity reinforces and stiffens an otherwise compliant amorphous polymer so that when oriented in fibers and films, semicrystalline polymers can exhibit outstanding physical properties. For example, highly oriented crystalline polyethylene fibers exhibit tensile properties nearly equivalent to those of carbon fibers, whereas the noncrystalline ethylene-propylene copolymers are soft, rubbery materials. Over the past 40 years, the semicrystalline state of (flexible) polymers has been the subject of many investigations, and some theoretical understanding has been generated concerning the thermodynamics and kinetics of crystallization. However, this area still has many unanswered questions. For example, the structures and properties of the crystal-amorphous interphase and their dependence on chemical structures need more attention in light of their technological importance for high-performance materials. Also needed is a better understanding of chain topology, such as entanglements and tie-chains interconnecting crystals, in the intercrystalline region, its variation with crystallization conditions, and its influence on bulk properties. Theoretical and simulation studies will aid greatly in further improving the properties of numerous synthetic fibers and films.

Liquid Crystalline Polymers

Rodlike polymers, and many semiflexible polymers with limited flexibility, are now known to exhibit liquid crystalline, or mesomorphic, order in solutions, in melts (the liquid state), and in the solid state. The spontaneous ordering of these polymers provides a unique means of aligning polymer chains to obtain materials having exceptional strength, rigidity, and toughness. Fibers spun from many of these polymers (Kevlar®, for example) exhibit mechanical properties that are almost equal to those of carbon fibers. Therefore, the fabrication of these polymers into fibers, films, and molded parts promises many new opportunities that have not been fully exploited because of a lack of understanding of the relationship of molecular structure to the properties of the liquid crystalline state. Theoretical advances will be key to establishing this badly needed understanding. They include describing the basic thermodynamics and kinetics of formation of the mesomorphic phase, the nature of defects (disclinations), the chain conformations and intermolecular packing in mesomorphic states, the dependence on processing history, and heterogeneities at large spatial scales. In particular, molecular descriptions are required that can explain and predict the influence of monomer structure, copolymer sequence, and interactions between chains.

Polymer Blends

Multicomponent polymer systems, such as polymer mixtures (blends), provide a new approach to the development of novel materials. These materials enable properties to be tailored, without resorting to costly new synthetic routes and also without the problems of proof of environmental and health safety entailed in new syntheses. Many important polymer properties, such as toughness, impact strength, heat and solvent resistance, and fatigue, have thus been improved significantly by blending polymers with different properties, and this has enabled many new commercial applications. Despite these advances, questions remain unanswered concerning predictions of miscibility and phase separation, control of morphology, and the structures of the interfaces between domains of different phases. Unlike small molecules, the mixing of two polymers is accompanied by only a small gain in entropy, and hence miscibility is more the exception than the rule. Because many of these materials must be processed in the liquid state, it is important to understand those features governing the phase diagrams of polymer liquid mixtures, that is, those factors determining whether the system is homogeneous or phase separated at a given temperature, pressure, composition, and so on. Traditionally, theories of this phase behavior have been based on simple models, but such models have not been fully satisfactory in relating the thermodynamic behavior of liquid polymer mixtures to the detailed chemical structures and interactions of their constituents; this is necessary for the molecular design of novel composite materials.

Newly emerging theories are establishing the much-needed link between phase behavior and molecular driving forces, through generalization of the classic lattice model of polymer solutions and through generalization, to polymers, of integral equation methods that have traditionally been applied to small-molecule liquids. A growing body of theoretical and experimental information suggests an alteration of chain dimensions in polymer blends, and further investigation is desirable. More difficult problems are those dealing with composite systems with mixtures of different phases, such as mixtures of crystalline with amorphous, liquid crystalline (stiff chains) with amorphous (flexible chains), and so on. For example, a challenging problem occurs when both amorphous phase separation and crystallization occur in the same system. There is coupling of the kinetics of the two processes, whereby the resulting morphology and consequent properties depend critically on thermal history. Similar problems exist to formulate molecular-level composites as mixtures of stiff and flexible chains. Very little theoretical work is available for addressing these problems and guiding critical experimental work.

Block Copolymers

To achieve a well-defined phase-separated morphology, in the process of designing materials with optimal properties, an important approach is to connect the two polymers of different chemical structures by covalent bonds, thereby producing block copolymers. Many prominent examples include thermoplastic elastomers used in commerce and biomedical technology. Block copolymers also have other uses, such as compatibilizing immiscible polymers, which arise from their ability to form micelles and self-assembled structures. Theories have begun to describe the morphologies of the self-assembled structures but are still unable to explain some observed structures. Further theoretical advances are needed to understand the features of their complex phase diagrams, such as the influence of chain architecture. Improved processing of materials containing block copolymers would benefit from theoretical guidance on the behavior of these systems in shear flows, in external fields, and under pressure, all of which are used as means for controlling morphology.

Polymer Interfaces

An increasing number of new composite materials are materials that have different phases dispersed throughout. The strengths of such materials are often determined by the strengths and morphologies of the interfaces between the phases. Adhesion is of intense interest: in some cases, such as for paints and bonding agents, adhesion should be high; in other cases, such as for lubricants, it should be low. Hence the need arises to understand the interfacial characteristics between different polymers and to understand the morphologies of compatibilizing agents, typically block or graft copolymers, that join two or more dissimilar components. Only recently have investigations considered the behavior of homopolymers and copolymers at interfaces and the nature of interfacial energies between polymers. There has been much progress in understanding properties of polymer interfaces, but we do not yet have a satisfactory understanding, at the molecular level, of the factors involved in strength and failure. In addition, the viability of certain polymeric materials is affected by the segregation of their components at the interfaces, so the properties of these interfaces are likewise of interest, especially in systems containing block copolymers. Theoretical and computational work will aid greatly in developing novel multicomponent polymer systems, where progress, although remarkable so far, still has a long way to go.

Polymer Surfaces and Thin Films

A number of surface properties of polymers, such as friction, wear, lubrication, adhesion, and sorption of surface species, have been investigated for many years because of their technological importance. But these studies have tended to be macroscopic and empirical, and thus little is known at the molecular level. It is observed that surface enrichment occurs where species preferentially concentrate at the surface according to composition and molecular weight. Surface segregation and phase separations near surfaces are not well understood. Nor is much known about the interactions between polymer surfaces and other materials such as liquid crystals, for example, in flat panel displays. Experimental methods to study the structures and properties of polymers on the 10- to 100-Å thickness scales are limited. Theoretical and computational efforts are critically needed to fill in details from the limited available experiments, which are often difficult and costly to perform.

Biopolymers

Biopolymers are another important class among polymeric states of matter. Biomolecules adopt an extremely wide variety of structures, spanning a large range of different types of organization and hierarchical complexity. Biology has control over specific monomer sequences, a power that is not yet available in synthetic polymer chemistry. It is the sequences, for example of proteins, RNA, and DNA, that control molecular architectures, and they do so with a high degree of precision. Force-field simulations for polymers were mainly developed first in the biomolecules area, and they continue to be of major importance in understanding biomolecule properties. More synergy is desirable between the polymer and biopolymer communities, because many of the needs and problems for theories and simulations are the same. Major needs in this area are for (1) theories and simulations that can couple and bridge a wide range of time and spatial scales and (2) better understanding of the complex interactions, such as electrostatic, hydrophobic, and hydrogen bond forces, that are important for biomolecules and other polymers in water.

Dynamics and Properties

Local Motions in Polymers

Localized motions in polymers include the same vibrational and torsional movements that are characteristic of motions in small molecules. However, the connectivity of a polymer chain introduces additional scales of "local motion," which can range from side-group rotations to cooperative movements involving segment sizes with tens of repeat units. It is these localized segmental motions that underlie variations in physical properties of polymers as a function of temperature and that influence mechanical and other properties of polymer materials. Theoretical descriptions of local motions lag far behind the experimental advances made in the last decade. Experimental advances have derived from tools such as NMR, fluorescence, electron spin resonance (ESR), and dielectric, rheo-optical, and dynamic mechanical instruments. Theoretical and simulation efforts are badly needed to provide a focus for producing a unified picture of chain dynamics with predictive capabilities based on fundamental principles. Also, currently available theoretical techniques have not yet linked the microscopic details of local motions to the long time relaxation behavior that is central to understanding mechanical, viscoelastic, and aging properties of materials. It is thus necessary to develop more molecular treatments of long time dynamics. Nowhere is this problem more apparent than in work on protein dynamics and folding, the separations of biomolecules according to their mobility under electric fields in polymer gels, and the rheology of entangled polymers. Important steps are being made in understanding how small molecules, such as gases and diffusants, affect local motions of polymeric chains, in order to diffuse through them. Such studies are necessary for developing better polymeric membranes for gas separations technology and better polymeric packaging materials for electronic circuitry, soft drink containers, and so on.

Rheology of Liquid Polymers

Molecular rheology has received considerable attention in the last 15 years. Substantial advances have been made by using the simple idea that the transport kinetics of polymers in dense media can be described in terms of the "reptating" motion of one chain within a medium of other chains by snakelike motions of the chain through a "tube" created by entangled neighboring chains. Much work remains to be done because current theories are still mainly of the single-chain type. New insights are required to better describe the nature of entanglement phenomena of the surrounding chains and to incorporate the relaxation of the surroundings and other motions on the rheological properties.

Advances in this area will also benefit from increased studies on copolymers in solvents, and on systems with complex topologies, such as stars and rings, and well-defined molecular weight distributions. We do not yet have a good theoretical understanding of the influence of chemical structure on rheology, because these effects are currently described by simple empirical monomer friction coefficients. Advances are likewise desirable in treating thermodynamics and phase behavior of polymers under the strong flows typical of processing conditions used to fabricate commercial materials. The molecular rheology of stiff chains, with its relationship to the nature of liquid crystalline order and their domain boundaries or disclinations, is another area that requires much more effort. An increased use of continuum rheological models to solve realistic polymer-processing flow problems has been spurred by the recent expansion in computing power. Further increases in computing power will enable workers to tackle more complex flows and to use the more sophisticated nonlinear models available. Instabilities in fibers, jets, wakes, and inlets will need continuing attention in this regard. Furthermore, a better connection between continuum and molecular rheology is necessary. Continuum rheology describes phenomena in terms of coefficients whose molecular significance is not clear. The rheology of multiphase systems is currently an active research area. This work considers the rheology due to assemblies of continuum phases with small dimensions. Examples are rigid fibers or more symmetrical ones in polymer fluids, phase-separated block or graft copolymers, and phase-separated polymer liquid crystals. Questions associated with polymer interfaces will be prominent in this area, because the wetting and bonding of the heterophases are important in these systems.

Mechanical Properties

There has recently been considerable improvement in molecular descriptions for the crazing and failure mechanisms of single-phase glassy polymers in terms of entanglements and chain breakage. The use of multiphase polymers, particularly block copolymers and polymer blends, has increased substantially in the last 10 years, but the deformation and toughening mechanisms of these systems are little understood. In general, a molecular-level description for mechanical properties of glassy polymers, such as toughness and fatigue, is still at a primitive stage, although computer simulation studies have recently shown promising results in understanding mechanical modulus and certain plastic flow processes.

Mechanical properties, such as adhesion, at polymer interfaces have recently received considerable theoretical attention, because they are critical to many technological problems. Present theoretical models that are based on simple single-chain-level pull-out and breakage need improvements in view of recent carefully designed experimental tests. All the highly oriented polymer fibers, such as Kevlar®, that exhibit outstanding tensile moduli and strength unfortunately show poor compressive strength, thus complicating their applications. Molecular-level description of compressive properties of oriented polymers is sorely needed in order to overcome this fundamental weakness in understanding high-strength polymers.

There is a continuing use of macroscopic nonlinear viscoelasticity models to describe polymer deformations, but the problems of uniqueness and physical significance of the representations are by no means solved. Fracture mechanics has proved useful in describing the failure of brittle materials and is now being extended to tougher materials. More work is required for the description of the failure process and its relation to other mechanical properties, especially for polymer matrix composites. Again, it is important to establish bridges between these continuum descriptions of fracture processes and molecular mechanisms involving stress concentrations on chemical bonds and the relationship to molecular topology.

Electro-active Properties

Conducting polymers have been an active research area during the last 15 years, but there are still fundamental problems in understanding the conduction processes. This understanding is critical to decreasing the bandgap of a conducting polymer to find an intrinsic conductor without any stability problems, to shifting the optical transitions out of the visible region in order to produce a transparent conductor, and to boosting the conductivity to obtain a truly metal-like conductor. Relatively little is known about the nature of conduction carriers for many of the polymers that are being investigated for this purpose. The mechanism of charge transport needs greater study, particularly with regard to the jumping of charge carriers between chains. Many unsolved problems persist concerning fabrication and stability of these materials, and theory could be very helpful in this regard.

Nonlinear optical properties of polymers, both second- and third-order nonlinearities, present enormous potential applications for future electro-optic devices, such as high-speed communications, data storage, and optical computing. Hence, the nonlinear optical properties exhibited by a large number of chromophore structures and polymer architectures have been investigated in the last several years. Despite the impressive amount of experimental work so far, advances in device applications are hampered by the lack of predictive capabilities for the optical absorption, linear optical, and nonlinear optical characteristics of chromophore structures, and chromophore-polymer interactions. Semiempirical computational methods, with the parameters optimized by using available experimental data, do provide reasonable trends within a given class of materials, but they are not reliably transferable to different types of chromophores. More theory is needed to relate these semiempirical methods to first-principle ab initio approaches to guide the improvement of the former methods. We need new theory dealing with methods for correlating many electrons.

Polymers have become an integral part of the current flat panel liquid crystal display technologies, where polymers form the aligning layers to give the liquid crystals the desired tilt angles. Moreover, polymer-dispersed liquid crystal systems show great promise for improved display applications. There is not yet theory for polymer-liquid crystal systems and their dynamics under electric fields. This is required to understand their display characteristics.

Computational Methods

As noted above, the growth in computing capabilities has enabled the modeling of polymer properties through computer simulations, often called computer experiments. Computer experiments are useful to address a wide range of questions and to model systems at the level of atomic interactions and in other situations at the macroscopic level. However, the simulation of equilibrium and dynamical polymer properties is complicated by the huge variations in processes that occur on disparate time and length scales.

Local small motions and small deformations occur on short time scales of picoseconds to nanoseconds while diffusion of entangled polymers and protein folding occur on time scales of milliseconds to seconds. Many other polymer processes occur on intermediate scales or over wide ranges of scales from the microscopic to the macroscopic. Different simulation methods (and theories) are needed for the different time and length scales.

Force-Field Simulation

One important computational tool is high-resolution force-field computer simulation. It is based on the interatomic forces, appropriately parameterized from either experiments on simple systems or ab initio quantum mechanical methods. Molecular dynamics (MD) methods, for example, use these forces to continually solve Newton's laws of motions for successive steps in time. A typical time step is about 1 femtosecond, at which the positions and velocities of the particles are recomputed. As the system evolves in time, the computer monitors the internal structures, the motions of molecules, and related properties. Given current capabilities, it is now possible to simulate systems having about 10,000 atoms to hundreds of picoseconds. These simulations are instructive for learning about the behavior of polymers and biopolymers and for capturing our understanding of them in underlying physical force laws on atomistic scales and in terms of the repeat chemical structure of polymers. However, the power of MD simulations is currently limited by the practically achievable time scales and physical sizes and the uncertainties in the force-field parameterization.

Figure 4.11 shows the power and limitations of force-field methods. The second line of the figure shows many processes of biopolymers, and the first line gives the time scale on which they occur. The third line shows processes of synthetic polymers. The power of force-field methods is the wide range of phenomena to which they can be applied, because all physical processes are ultimately derivable from the underlying forces. The main limitation of the method is that computer power is not currently great enough to explore many properties. The fourth line shows projections for the year in which various properties are expected to be accessible by simulations. This time line is based on estimated developments in parallel-processing computing. The maximum run length for a small protein in water in 1992 was about 500 picoseconds (ps), which requires a few hundred hours on a supercomputer. But a single run does not yield an understanding of a process. In 500 ps, we can observe about 10 occurrences of a process that has a 50-ps relaxation time. If we need 10 different samples of the computer experiment in order to understand the statistical errors,

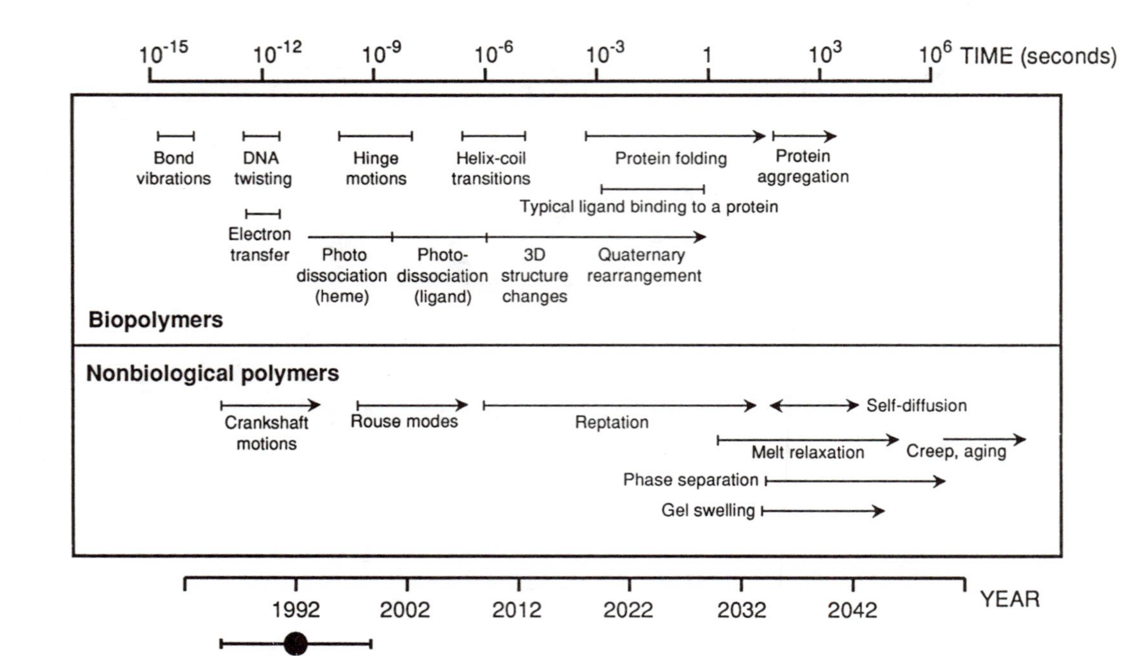

FIGURE 4.11 Time scales for various motions within biopolymers and nonbiological polymers. The year scale at the bottom shows estimates of when each such process might be accessible to force-field simulation on supercomputers, assuming that parallel processing capability on supercomputers increases at about the rate of 10^3 every 10 years and neglecting new approaches or breakthroughs. At current capabilities, a given allotment of computer time can be used for one run performed over a few hundred picoseconds for a small protein in a few thousand water molecules, or for 1,000 runs to explore 1,000 processes that have relaxation times of hundreds of femtoseconds; this range is indicated by the error bar below the year scale. SOURCE: Reprinted with permission from Chan and Dill (1993), p. 29. Copyright © 1993 by the American Physical Society.

then our "understanding" of processes could be said to be only in the tens of picoseconds range at the present time. By these estimates, it will be tens of years until many processes such as protein folding, glass transition, shear thinning, and creep and aging in polymers will be fully understood by force-field simulation.

Nevertheless, important steps are already being made. For example, it is expected that advances in algorithms that exploit newer techniques to integrate Newton's equations of motion for the system will be developed to effectively increase the size of the time step. Among these methods are united-atom algorithms, Monte Carlo sampling of configuration space coupled with MD, Brownian dynamics, and the stiff-equation methods. By these means, times into the nanosecond real-time domain will become accessible. One of the most important aspects of MD simulation is the fact that the algorithms are well suited to vectorization and parallelization. With the development of teraflop machines, MD codes will be among the first to be developed for massively parallel computers. The machines will be approximately 1,000 times faster than the current fast single processors, and the time domains that can be reached will then be on the order of microseconds of real time. While this is not yet into the very long times of seconds to years that characterize the longer relaxation times in polymers, the dynamical range that will be covered with the new technology could span 9 orders of magnitude (from femtoseconds to microseconds).

Continued progress will also require improvements and stringent testing in force-field potentials and parameters. Advances in computational capabilities will allow the ab initio quantum mechanical calculations of geometries and force-field energies to be carried out on appropriate chain segment scales with adequate basis functions and rudimentary electron correlation effects. However, limitations on treating electron correlation put the reliable accuracy in these potentials as optimistically no better than about 0.3 kilocalories per mole, nearly comparable to RT (where R is the gas constant and T is the temperature). Hence, the results of quantum mechanical calculations will provide a good guide to force-field parameterization, but the parameters must be adjusted further and tested against experiments.

Coarse-grained Simulations

An alternative to solving Newton's laws for the dynamics of molecules is the use of statistical sampling methods, referred to as Monte Carlo, or complete enumeration of polymer or biopolymer conformations. These approaches are often used with lower-resolution representations of chain molecules (e.g., where chains are represented as strings of "beads," rather than in atomic detail). The advantage over high-resolution simulations is the ability to explore more conformations and thereby represent more accurately some physical properties. Exhaustive simulations are the most complete in this regard, but they suffer either a power-law or exponential dependence of computer time on chain length and are therefore not practical for long chains. Nevertheless, exhaustive simulations and Monte Carlo methods have been important for testing physical principles, for testing underlying assumptions in other simple models of polymer behavior, and for motivating some of the theoretical developments described below.

Among the most exciting new developments are the Gibbs ensemble Monte Carlo and configurational bias Monte Carlo techniques. They permit computation of phase diagrams from force-field simulations, and chemical potentials for insertion of polymer chains, even into high-density media.

More advances in sampling methods and in algorithms are needed, especially to take advantage of massively parallel computer architectures. Such advances would be particularly valuable for improving our abilities to simulate dense polymer systems and thereby guide further experiments and theoretical developments.

A lower-resolution alternative to force-field simulation for dynamic properties is known as the Brownian dynamics simulation. This method usually models the polymer chain as beads interconnected by elastic springs and moving in a viscous medium. While this method allows the probing of much longer time scales, the spring and viscous damping constants are not directly derivable from atomic quantities, and the model itself becomes problematic for dense polymer systems. Better algorithms and models are required, as well as statistical mechanical methods that will link the parameters in the Brownian dynamics method to molecular-level potentials available in the force-field simulations.

Impressive advances are also being made in the atomistic modeling of glassy polymers. In this approach, the model system is a cube with periodic boundaries, filled with polymer segments. An initial configuration is generated and then relaxed by potential energy minimization. The resulting "equilibrium" structure is examined for predictive and interpretive information. In this way, it is possible to obtain estimates of cohesive energy densities, Hildebrand solubility parameters, degree of randomization of the "amorphous" chains, elastic constants, thermal expansion coefficients, and structural changes brought about by shear and tension deformations. One of the current directions in which these simulations are being extended is to predictions of chain dispositions in the vicinity of a bounding surface. Another is the diffusion of small molecules through polymer films, which is relevant to the understanding of gas-separation membranes and the design of materials of improved permeability and selectivity. Calculations such as these are proving to be quite successful in predicting and interpreting polymer properties and should be extended to other applications.

Another centerpiece of polymer science is the rotational isomeric state (RIS) theory developed in Russia, Japan, Israel, and the United States in the 1950s. For certain conditions, it predicts a wide range of polymer properties from the known chemical constitution of the monomer units. Its premise is that skeletal bonds are in one or another of a few favored conformations. Its power comes from its separation of the computation of a polymer conformation into local and nonlocal aspects. Computer resources can be used efficiently because the main computational effort is directed only toward extracting information about two-bond or three-bond units, rather than being expended inefficiently in exploring conformations of a long chain. The statistical mechanical machinery then uses that information efficiently to predict the properties of long chains. The RIS theory is a paradigm for combining molecular simulations with statistical mechanics in a way that makes most effective use of both methodologies. Applications to date have focused on isolated chains, and most involve the calculation of equilibrium properties such as chain dimensions, but recent work addresses applications to dynamic properties as well.

Mathematical Methods

With the tremendous increase in computing capability, it will become feasible to solve many fundamental statistical mechanical equations by numerical integrodifferential methods. Recent advances in ab initio quantum mechanical methods, for example, have been made by advanced numerical methodologies that take advantage of the processing power of present supercomputers or advanced workstations. Similar advances in mathematical methodologies will turn out to be as useful for many polymer problems.

Conclusions

- Theoretical and computational methods play a fundamental role in developing new properties and polymeric materials. Explosive developments in computer technology have fueled the growth of computational experiments and simulations. But computational studies should not overshadow the development of deeper and more rigorous theory, because theory is a powerful wellspring of major paradigm changes. Unique insights will come from theorists and physicists, who should be encouraged to explore polymers and biomolecules.
- Computational and theoretical methods are often limited to treating narrow ranges of time and spatial scales. But polymers and biomolecules have behaviors that can range over tens of orders of magnitude in time and space within a single system. An important area of future development is methods that can bridge broadly from the fast and microscopic scales to the slow and macroscopic scales. We need to relate continuum fracture mechanics models to atomic levels of bonding, semiempirical predictions of optical properties to ab initio quantum mechanics, picosecond atomic motions of proteins to tens-of-seconds folding processes, atomic bond rotations and interactions to macroscopic glass transitions, and so on. Totally new modeling approaches are needed to bridge these gaps.
- Force-field simulations and experiments have grown enormously. Further developments will rely not only on increasing computer power and increased

access to supercomputers and to fast networks, but also on improved force fields, deeper connections to quantum mechanics, and better treatments of the environment of surrounding and entangled chains or solvents. Better quantum mechanical methods are needed to treat large numbers of electrons and atoms.

- Practical challenges include applying theory and computation to polymer behaviors in complex media—polymer blends, liquid crystalline polymers, semicrystalline materials, composites, block copolymers, interfaces, the rheology of mixtures, branched molecules, soft matter, and so on.

REFERENCES

Butera, R., L.J. Fetters, J.S. Huang, D. Richter, W. Pyckout-Hinzen, A. Zirkel, B. Farago, and B. Ewen. 1991. "Microscopic and Macroscopic Evaluation of Fundamental Facets of the Entanglement Concept." *Physical Review Letters* 66:2088.

Chan, H.S., and K.A. Dill. 1993. "The Protein Folding Problem." *Physics Today* 46(2):24-32.

Gibson, H.W., C. Wu, Y.X. Shen, M. Bheda, A. Prasad, H. Marand, E. Marand, and D. Keith. 1992. "Synthesis, Thermal and Phase-Behavior of Polyester, Polyurethane and Polyamide Rotaxanes." *Polymer Preprints* 33(1):235.

Mesei, F. 1980. *Lecture Notes on Physics*. Vol. 122. Berlin: Springer Verlag.

Richter, D., B. Farago, L.J. Fetters, J.S. Huang, B. Ewen, and C. Lartique. 1990. "Direct Microscopic Observation of the Entanglement Distance in a Polymer Melt." *Physical Review Letters* 64:1389.

Tomalia, D.A., D.R. Swanson, and D.M. Hedstrand. 1992. "Comb-burst Dendrimers—A New Macromolecular Architecture." *Polymer Preprints* 33(1):180.

APPENDIX

Contributors and Participants

WRITING CONTRIBUTORS

Gibson L. Batch, 3M
Lee Landis Blyler, Jr., AT&T Bell Laboratories
Martin C. Cornell, The Dow Chemical Company
David F. Eaton, E.I. du Pont de Nemours & Co.
Bruce E. Eichinger, BioSym Technologies
Karl F. Freed, University of Chicago
John L. Gardon, AKZO Coatings Am-ICA
Russell Gaudiana, Polaroid
Steve Goldman, Proctor and Gamble Company
William W. Graessley, Princeton University
Werner Grootaert, 3M
James E. Guillet, University of Toronto
Edward G. Howard, E.I. du Pont de Nemours & Co.
Richard Ingwall, Polaroid
Michael Jaffe, Hoechst Celanese Corporation
Peter C. Juliano, General Electric Company
Robert S. Langer, Massachusetts Institute of Technology
Mel A. Leitheiser, 3M
Andrew Lovinger, AT&T Bell Laboratories
Alan G. MacDiarmid, University of Pennsylvania
L.T. Manzione, AT&T Bell Laboratories
P.A. Mirau, AT&T Bell Laboratories

James Moore, Rensselaer Polytechnic Institute
Murugappan Muthukumar, University of Massachusetts
Allen Noshay, Union Carbide Corporation
Alphonsus V. Pocius, 3M
John F. Rabolt, IBM Almaden Research Center
Kenneth L. Reifsnider, Virginia Polytechnic Institute and State University
Thomas P. Russell, IBM Almaden Research Center
Felix Theeuwes, ALZA Corporation
Garth L. Wilkes, Virginia Polytechnic Institute and State University
David J. Williams, Eastman Kodak Company
Gary E. Wnek, Rensselaer Polytechnic Institute
Fred Wudl, University of California at Santa Barbara
Do Y. Yoon, IBM Almaden Research Center
Bruno H. Zimm, University of California at San Diego

QUESTIONNAIRE PARTICIPANTS

R. Stephen Berry, University of Chicago
Paul Calvert, University of Arizona
John G. Curro, Sandia National Laboratory
Alan D. English, E.I. du Pont de Nemours & Co.
Glenn H. Fredrickson, University of California at Santa Barbara
Peter C. Juliano, General Electric Company
Wayne L. Mattice, University of Akron
Robert M. Nowak, Dow Chemical Company
Eli M. Pearce, Polymer Research Institute, Polytechnic University, Brooklyn
Philip Pincus, University of California at Santa Barbara
Hans Pohlmann, Amoco Chemical Corporation
Durward T. Roberts, Bridgestone/Firestone
Ann Salamone, Rochal Industries
Edward T. Samulski, University of North Carolina at Chapel Hill
George Schmeltzer, Miles
Matthew V. Tirrell, University of Minnesota
S. Richard Turner, Eastman Kodak Company
Garth L. Wilkes, Virginia Polytechnic Institute and State University
Tom Wollner, 3M
Hyuk Yu, University of Wisconsin at Madison

WORKSHOP PARTICIPANTS

Charles E. Browning, Wright Laboratory, Wright-Patterson Air Force Base
Stuart L. Cooper, University of Wisconsin at Madison
Kenneth A. Dill, University of California at San Francisco

Andrew E. Feiring, E.I. du Pont de Nemours & Co.
William W. Graessley, Princeton University
James E. Guillet, University of Toronto
Melvin A. Leitheiser, 3M
Bruce M. Novak, University of California at Berkeley
Virgil Percec, Case Western Reserve University
John F. Rabolt, IBM Almaden Research Center
Steven D. Smith, Proctor and Gamble Company
Edwin L. Thomas, Massachusetts Institute of Technology
George M. Whitesides, Harvard University
Joseph G. Wirth, Raychem Corporation
Martel Zeldin, Indiana University and Purdue University at Indianapolis

Index

H

I

K

L

M

N

O

P

R

U

V

W

X

Z